Interaction for Visualization

Synthesis Lectures on Visualization

Editor
Niklas Elmqvist, *University of Maryland*
David S. Ebert, *Purdue University*

Synthesis Lectures on Visualization publishes 50- to 100-page publications on topics pertaining to scientific visualization, information visualization, and visual analytics. Potential topics include, but are not limited to: scientific, information, and medical visualization; visual analytics, applications of visualization and analysis; mathematical foundations of visualization and analytics; interaction, cognition, and perception related to visualization and analytics; data integration, analysis, and visualization; new applications of visualization and analysis; knowledge discovery management and representation; systems, and evaluation; distributed and collaborative visualization and analysis.

Interaction for Visualization
Christian Tominski
2015

Data Representations, Transformations, and Statistics for Visual Reasoning
Ross Maciejewski
2011

A Guide to Visual Multi-Level Interface Design From Synthesis of Empirical Study Evidence
Heidi Lam and Tamara Munzner
2010

Interaction for Visualization

Christian Tominski

ISBN: 978-3-031-01472-7   paperback
ISBN: 978-3-031-02600-3   ebook
ISBN: 978-3-031-03728-3   ePub

DOI 10.1007/978-3-031-02600-3

A Publication in the Springer series
SYNTHESIS LECTURES ON VISUALIZATION

Lecture #3
Series Editors: Niklas Elmqvist, *University of Maryland*
                David S. Ebert, *Purdue University*
Series ISSN
Print 2159-516X   Electronic 2159-5178

Interaction for Visualization

Christian Tominski
University of Rostock

SYNTHESIS LECTURES ON VISUALIZATION #3

ABSTRACT

Visualization has become a valuable means for data exploration and analysis. Interactive visualization combines expressive graphical representations and effective user interaction. Although interaction is an important component of visualization approaches, much of the visualization literature tends to pay more attention to the graphical representation than to interaction.

The goal of this work is to strengthen the interaction side of visualization. Based on a brief review of general aspects of interaction, we develop an interaction-oriented view on visualization. This view comprises five key aspects: the data, the tasks, the technology, the human, as well as the implementation. Picking up these aspects individually, we elaborate several interaction methods for visualization. We introduce a multi-threading architecture for efficient interactive exploration. We present interaction techniques for different types of data (e.g., multivariate data, spatio-temporal data, graphs) and different visualization tasks (e.g., exploratory navigation, visual comparison, visual editing). With respect to technology, we illustrate approaches that utilize modern interaction modalities (e.g., touch, tangibles, proxemics) as well as classic ones. While the human is important throughout this work, we also consider automatic methods to assist the interactive part.

In addition to solutions for individual problems, a major contribution of this work is the overarching view of interaction in visualization as a whole. This includes a critical discussion of interaction, the identification of links between the key aspects of interaction, and the formulation of research topics for future work with a focus on interaction.

KEYWORDS

Visualization, interaction, zoomable user interfaces, interactive lenses, touch interaction, tangible interaction, proxemic interaction, selection, navigation, exploration, visual comparison, visual editing, large displays, automatic methods, navigation recommendations

Contents

Acknowledgments ix

1 Introduction 1
    1.1   Why Focus on Interaction? 1
    1.2   An Interaction-oriented View 2
    1.3   Outline 3

2 Fundamentals 5
    2.1   Visualization 5
    2.2   Human–Computer Interaction 7
    2.3   Interaction in Visualization 9
    2.4   The Visualization–Interaction Gap 17
    2.5   Interaction—Useful or Harmful? 19
    2.6   Implementing Interactive Visualization 21
    2.7   Summarizing Thoughts 23

3 Aspects of Interaction in Visualization 25
    3.1   The Data 26
    3.2   The Tasks 27
    3.3   The Technology 28
    3.4   The Human 29
    3.5   The Implementation 30

4 Methods and Techniques for Interactive Visualization 33
    4.1   An Architecture for Efficient Interactive Visualization 34
    4.2   Data Characteristics and Interaction 36
            4.2.1 Interacting with Graphs 37
            4.2.2 Interacting with Spatio-temporal Movement Trajectories 40
    4.3   Task-specific Interaction Techniques 45
            4.3.1 Interaction for Comparison Tasks 45
            4.3.2 Interaction Support for Editing Tasks 48

4.4 Utilizing Modern Technology for Interaction 51
 4.4.1 Tangible Views for Interaction and Visualization 52
 4.4.2 Proxemic Interaction for Wall-sized Visualization 55
4.5 Supporting Human Interaction with Automatic Methods 59
 4.5.1 Reducing Interaction with Event-based Concepts 60
 4.5.2 Navigation Recommendations for Informed Interaction 62
4.6 Summarizing Remarks 66

5 Conclusion and Future Work **69**
5.1 Concluding Remarks 69
5.2 Topics for Future Work 71

Bibliography **75**

Author's Biography **97**

Acknowledgments

This work would not have been possible without the collaboration and support of many people. I'm deeply grateful for being given the opportunity to know them, to learn from them, and to work with them.

First and foremost, I thank Heidrun "Heidi" Schumann for being a great mentor, for maintaining an unparalleled working environment, for providing invaluable advice and motivation, for always seeing things positively, and simply for being a dear friend!

I thank Niklas Elmqvist and David Ebert for giving me the opportunity to publish in their synthesis lecture series on visualization. Diane Cerra did a great job in managing this lecture at Morgan & Claypool.

I'm deeply grateful to Heidrun "Heidi" Schumann, Jarke van Wijk, Helwig Hauser, Raimund Dachselt, and Hans-Jörg "Hansi" Schulz, who provided valuable feedback on an early draft of this work. I'd like to extend my warmest thanks to Petra Isenberg and Niklas Elmqvist for reviewing the present version.

This lecture builds upon publications that I co-authored with several fellow researchers. James Abello worked with me on the interactive graph visualization system CGV. I had the pleasure to work with Harald Piringer and his colleagues on multi-threading for visualization. The collaboration on tangible views with Martin Spindler and Raimund Dachselt opened my view on what lies beyond mouse and keyboard. In a similar way, the collaboration with Anke Lehmann drew my attention from regular desktop displays to large high-resolution display walls. Camilla Forsell and Jimmy Johansson were of great help in studying interaction for visual comparison. I thank Natalia and Gennady Andrienko for collaborating with me on visualizing movement trajectories. Stefan Gladisch deserves many thanks for working with me on editing graph structures and on navigation recommendations. To all my collaborators, thank you, I hope we do some joint work again soon.

Finally, I thank my family! Anja, Vincent, and Arvid, I dedicate this work to you!

Christian Tominski
June 2015

CHAPTER 1

Introduction

Nowadays, we live in a world full of data. Technological advances have led to a situation where we collect excessively far more data than we can make sense of—a problem known as *information overload* (Strother et al., 2012). *Visualization* has become an accepted means to address the information overload. The key idea behind visualization is to transform data into pictures that humans can interpret more easily than large quantities of numbers (Ware, 2012). *Interaction* between the human and the computer plays an integral role in the process of forming mental models of the data (Spence, 2007). This work emphasizes the role of interaction in visualization.

1.1 WHY FOCUS ON INTERACTION?

Bertin (1981) points out the importance of interaction for visual data exploration and analysis:

> "A graphic is not 'drawn' once and for all; it is 'constructed' and reconstructed until it reveals all the relationships constituted by the interplay of the data. The best graphic operations are those carried out by the decision-maker himself."
>
> — Bertin (1981)

Bertin conveys two key messages. First, interaction is indispensable for constructive processes such as developing insight into complex data, and second, interaction enables the human to steer the data exploration and to make the final assessment of the data. Interestingly, Bertin expressed his thoughts on interaction years before visualization existed as a field. Still, the essence of his statement remains valid until today.

Computer-supported visualization has always included the notion of interactivity. Similar to what Bertin said, Pike et al. (2009) state the following:

> "It is through the interactive manipulation of a visual interface–the analytic discourse–that knowledge is constructed, tested, refined and shared."
>
> — Pike et al. (2009)

Despite the importance of interaction in visualization, much of the literature on visualization focuses, in fact, on the visual part, not so much on the interaction part. Many visualization publications describe in detail aspects of the visual representation, but less is reported about the design and the implementation of interaction in visualization. Several other researchers have taken note of this deficiency:

"Even though interaction is an important part of information visualization (Infovis), it has garnered a relatively low level of attention from the Infovis community."

— Yi et al. (2007)

"Until recently, the focus of InfoVis has been more on the graphical representation and less on the interaction."

— Fekete (2010)

"Also, although interaction isn't yet a primary theme, the visualization research literature reflects an increasing focus on it."

— Keefe (2010)

"Unfortunately, interaction is not discussed at all in graphic design, and even visualization textbooks tend to downplay this angle."

— Elmqvist et al. (2011)

There are several explanations for why interactive aspects are not on equal terms with visual aspects. First, visualization has its roots in computer graphics. Second, reporting a result related to interaction (a process) is typically more difficult than reporting on a visual result (an image). Third, there is no standard notation one could rely on when describing interaction.

This work is an attempt to balance the visualization plus interaction equation. To this end, we deliberately look at visualization from an interaction perspective.

1.2   AN INTERACTION-ORIENTED VIEW

We develop and discuss an interaction-oriented view of visualization, bringing together the relevant concerns under a common hood. *Data* and *tasks* are key factors of visualization and likewise they are primary concerns to be considered for interaction. Further, we consider the *technology* providing the means for display, interaction, and computation, as well as the *human* user as the recipient of visual information and active participant in the interactive data exploration and analysis process.

That said, the primary topic of interest of this work is to investigate interaction in visualization along the key factors: data, tasks, technology, and human. Studying these factors individually with a focus on interactive approaches, we provide a broader picture on interaction in visualization as a whole. Addressing the *data*, we discuss solutions taking into account both the structure of data as well as the spatial and temporal frame of reference in which data are usually given. With regard to *tasks*, we present interaction techniques for visual comparison and data editing, both of which being highly relevant in data-intensive work places. We introduce techniques that utilize different interaction *technologies*, including classic mouse and keyboard interaction, but also modern touch interfaces and physical interaction in front of large high-resolution displays. Focusing on the *human* user, we look at techniques for reducing interaction costs by drawing inspiration from natural interaction, by following real-world workflows, and by integrating automatic methods.

The interaction side of visualization is also studied from an *implementation* perspective. We present an efficient multi-threading architecture that can serve as a general basis for developing interactive visualization systems. We further illustrate several solutions that incorporate modern display technology and interaction modalities to implement novel ways of interacting with visual representations of data.

With our interaction-oriented view organized according to data, tasks, technology, human, and implementation, we hope to contribute to lifting interaction in visualization to a level that corresponds to its widely acknowledged importance.

1.3   OUTLINE

Chapter 2 starts with an introduction to the fundamental concepts of visualization and interaction. The introduction collects various definitions, explains basic interaction techniques, studies the visualization-interaction gap, and discusses the advantages and disadvantages of interaction in visualization. Basics of implementing interactive visualization solutions complement this chapter.

Chapter 3 takes a closer look at the aspects of interaction in visualization and develops a structured interaction-oriented view on the topic. As indicated, we will cover five key aspects: the data, the tasks, the technology, the human, and the implementation.

In Chapter 4, we present methods and techniques that illustrate key questions and corresponding solutions with respect to our interaction-oriented view. In Section 4.1, we address interaction implementation on a fundamental level by discussing a multi-threading architecture for interactive visualization applications. Section 4.2 sets the focus on the data aspect by introducing effective ways of interacting with graph structures and movement trajectories in space and time. The task aspect is taken up in Section 4.3, where we illustrate interaction techniques for visual data comparison and data editing tasks. In Section 4.4, we present tangible views and explain physical navigation in front of large displays as novel ways of interaction that take advantage of technological progress. Addressing the human user, we discuss the use of automatic event-based methods and navigation recommendations as means to reduce interaction costs in Section 4.5. All approaches are described in a compact way, presenting the key messages with a focus on interaction in visualization.

Chapter 5 provides an overall summary and conclusion. Key concerns are to derive and discuss insights about the greater picture of interaction in visualization as drawn in this work and to identify research topics for future work.

CHAPTER 2

Fundamentals

This chapter takes a look at some fundamental aspects of visualization and interaction. We will first consider visualization and human-computer interaction in general, before we shift our focus to visualization-specific questions of interaction.

2.1 VISUALIZATION

As early as in the 1980s, visualization pioneers recognized the enormous potential that modern computers would offer in terms of analytic power, graphics output, and interactive manipulation. McCormick et al. (1987) formulate the key idea of visualization as:

> "Visualization is a method of computing. It transforms the symbolic into the geometric, enabling researchers to observe their simulations and computations. Visualization offers a method for seeing the unseen. It enriches the process of scientific discovery and fosters profound and unexpected insights."
>
> — McCormick et al. (1987)

With their seminal work, McCormick and colleagues paved the way for visualization as a distinct field of computer science. About a decade later, Card et al. (1999) define visualization—the visualization of information in particular—as follows:

> "Information visualization: The use of computer-supported, interactive, visual representations of data to amplify cognition."
>
> — Card et al. (1999)

At its core, the definition brings together the capabilities of human perception and cognition and the computational abilities of computers as the key components glued together by interaction. Ware (2008) further emphasizes the interplay of the human and the computer:

> "It is useful to think of the human and the computer together as a single cognitive entity, with the computer functioning as a kind of cognitive coprocessor to the human brain. [...] Each part of the system is doing what it does best. The computer can preprocess vast amounts of information. The human can do rapid pattern analysis and flexible decision making."
>
> — Ware (2008)

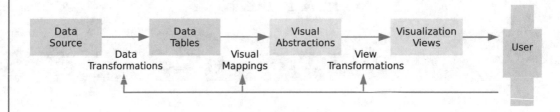

Figure 2.1: The visualization pipeline (adapted from Card et al., 1999).

For the interplay of human and computer to be successful, the visualization literature describes a number of criteria that must be observed when data are transformed into visual representations. The economic model of visualization by van Wijk (2006) demands that the benefits of using visualization as a means to gain insight outweigh the costs involved in carrying out the process (i.e., computation and interpretation). A necessary condition to achieve beneficial visual representations is to consider the two key visualization criteria, *expressiveness* and *effectiveness*, as identified by Mackinlay (1986). Visual representations are expressive if they actually do express the desired information. Effectiveness relates to making the most of the capabilities of the human visual system and the output medium.

Striving to meet the expressiveness and effectiveness criteria, visualization approaches typically follow the *visualization pipeline* model. The pipeline model is useful not only for describing and implementing the process of generating visual representations of data (Haber and McNabb, 1990; Chi, 2000; dos Santos and Brodlie, 2004) but also for explaining the user's involvement in this process (Card et al., 1999; Jansen and Dragicevic, 2013).

The visualization pipeline by Card et al. (1999) is a widely accepted blueprint for visualization. As depicted in Figure 2.1, the pipeline prescribes, at a most abstract level, how data is to be transformed through several stages from a data source into data tables, into visual abstractions, and finally into visualization views. Data transformations, including filtering, clustering, error correction, can be found in the early stage of the pipeline.

At the heart of the pipeline are the visual mappings, that is, the transformation of data tables to visual abstractions (i.e., graphical primitives). Bertin (1983), Mackinlay (1986), and MacEachren (1994) describe different *visual variables* that can bear information in a visual representation. Examples are given in Figure 2.2, including position, size, shape, orientation, hue, saturation, brightness, texture, focus, transparency, containment, and connection. Different visual variables have different expressive power and work differently well for different types of data. The choice of visual variables largely determines how effective a visual representation is. Mackinlay (1986) and, more recently, Heer and Bostock (2010) suggest suitable mappings of data to visual variables. For example, size is good for numeric data, whereas shape is more suitable for categorical data.

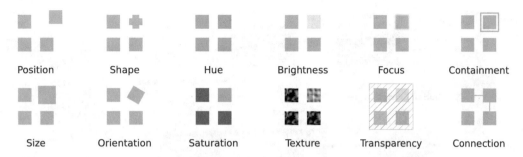

| Position | Shape | Hue | Brightness | Focus | Containment |

Figure 2.2: Visual variables for encoding data graphically.

Finally, the generated graphics are rendered to visualization views, that is, to visual representations. At the view transformation stage, typical computer graphics operations are carried out, including projection, clipping, and rasterization.

Let us now consider how users are involved in the visualization process. The particular visualization pipeline in Figure 2.1 incorporates the user in two different roles. On the one hand, the user is the recipient of the information communicated via visualization views. In this role, the user aims to make sense of visual representations and to link features in the graphics to actual characteristics of the underlying data. On the other hand, the pipeline suggests that the user can be an active participant controlling the different stages of the transformation from data to views. Unfortunately, the pipeline model does not provide the same level of detail for generating visual representations and for interactively working with them—reason enough for us to look more closely at interaction.

2.2   HUMAN–COMPUTER INTERACTION

A primary source of scholarly literature on interaction is the realm of human–computer interaction (HCI). According to Dix et al. (2004), "HCI involves the design, implementation and evaluation of interactive systems in the context of the user's task and work."

A general model for interaction is described by Norman (1988, 2013). According to this model, interaction can be conceptualized as a cycle of two phases: the execution phase and the evaluation phase. As both phases occasion costs, Norman denotes them as gulfs: the *gulf of execution* and the *gulf of evaluation*.

In the context of interaction between humans and computers, Norman's model can be illustrated as in Figure 2.3. Starting with a goal (e.g., read web page), the execution phase is concerned with the intent to interact (e.g., scroll down), the mental planning of the interaction (e.g., plan to use mouse wheel), and the actual execution of the plan (e.g., rotate mouse wheel). Performing this first phase of the action cycle results in a response from the system (e.g., redisplay web page). The evaluation of the response is captured in the second phase of the action cycle. This includes

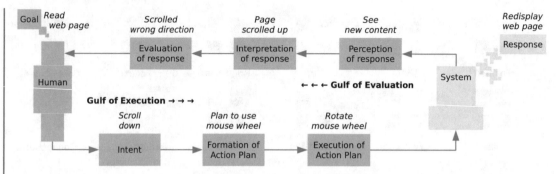

Figure 2.3: Interaction as a cycle of execution and evaluation phases.

the perception of the response (e.g., see new content), the mental interpretation of it (e.g., page scrolled up), and its actual evaluation with regard to the intent that induced the interaction (e.g., scrolled wrong direction). If outcome and intent of the interaction do not match, the cycle is re-run—certainly with the same intent, but with a different plan (e.g., use touch to scroll) that hopefully leads to a positive evaluation.

Similar to expressiveness and effectiveness in visualization, there are certain criteria that interaction has to obey. *Usability* (Nielsen, 1993) and *user experience* (Hassenzahl and Tractinsky, 2006) are key aspects in this regard. They subsume criteria such as predictability, consistency, customizability, satisfaction, engagement, responsiveness, and task conformance, to name only a few. The rationales behind these criteria are standardized in ISO standards 9241-11 on usability and 9241-210 on human-centered design.

For the practitioner, the literature also provides rules and guidelines. Shneiderman and Plaisant (2009) list the following eight *golden rules*:

1. Strive for consistency
2. Cater to universal usability
3. Offer informative feedback
4. Design dialogs to yield closure

5. Prevent errors
6. Permit easy reversal of actions
7. Support internal locus of control
8. Reduce short-term memory load

Norman (2013) describes seven fundamental principles of design. According to these principles, it must be possible to determine the state of a system and possible actions at all times (*discoverability*). Appropriate feedback reflects the response to an action and the new system state (*feedback*). The relevant information is presented in a way that leads to understanding and a feeling of control (*conceptual model*). Suitable options are available to carry out the intended actions (*affordances*). Well perceivable indicators support discoverability and feedback (*signifiers*). Controls naturally map to the actions they trigger (*mappings*). Constraining the available actions can offer guidance and ease interpretation (*constraints*).

The criteria we included here are not exhaustive. The reader is referred to the HCI literature for a more comprehensive discussion. Dix et al. (2004) and Shneiderman and Plaisant (2009) dedicate entire chapters to principles, rules, guidelines, and best practices.

Striving for effective and efficient action-response cycles, HCI researchers have proposed several themes of interaction. Early research focused on command-line interfaces and interaction using windows, icons, menus, and pointers, commonly known as the WIMP paradigm (Dix et al., 2004). A classic and most prevalent theme, also in visualization, is *direct manipulation* as proposed by Shneiderman (1983). Direct manipulation can be characterized as interaction with constantly visible objects that are manipulated by physical and reversible actions with immediate feedback. Building upon direct manipulation, Beaudouin-Lafon (2000) coins the term *instrumental interaction*. It captures the idea of using interaction instruments as mediators to manipulate domain objects.

With the advent of modern interaction devices, new ways of interacting became possible. It began the so-called post-WIMP era (van Dam, 1997). *Tangible interaction* (Ishii and Ullmer, 1997) is based on interaction with tangible objects in the real world. *Reality-based interaction* (Jacob et al., 2008) and *natural interaction* (Valli, 2008) are the next steps in a line of recent developments that include aspects of greater awareness of the user and the environment in which the interaction takes place.

These models and studies from the HCI literature contribute largely to a better understanding of the role and needs of human beings interacting with computers in general. The specific aspects of interaction in the context of visualization deserve further elaboration.

2.3   INTERACTION IN VISUALIZATION

Spence (2007) describes visualization as a tool to support humans in forming mental models of otherwise difficult-to-grasp complex phenomena. The fact that people *form* mental models suggests that interaction be a principal ingredient of visualization. In fact, the visualization pipeline by Card et al. (1999) (see Figure 2.1 on page 6) includes interaction by means of a human, a user who actively controls the individual transformation stages and receives corresponding visual feedback.

There are visualization scenarios that get along well with only minimal interaction or no interaction at all. An example is information graphics for which the data are usually understood quite well and the goal is to communicate results or facts about the data.

Yet, when the visualization goal is of open-ended exploratory nature, interaction becomes indispensable. Especially when dealing with unknown data, their visual representations might not be as expected due to the manifold factors that influence the visualization process (e.g., data characteristics, analytic tasks, choice and parametrization of analytical and visual methods). Unfortunately, we usually do not know exactly what the expected visual outcome is or whether it is effective with respect to the data and the tasks at hand. Because there are things we do not know, we have to seek assistance from the user.

It becomes clear that visualization is not a one-way street of transforming data into images, but in fact can be considered a human-in-the-loop process of controlled transformations (Conversy, 2013). Interaction helps people in understanding the visual encoding, in realizing the effect of visualization parameters, in identifying key data characteristics and carving out hidden patterns, and eventually, in becoming confident about the data.

Interaction also provokes curiosity. Analysts want to get their hands on their data, and interaction enables them to experiment with different *what-if* scenarios. The importance of interaction in visualization is nicely reflected in a statement by Thomas and Cook (2005):

> "Visual representations alone cannot satisfy analytical needs. Interaction techniques are required to support the dialogue between the analyst and the data."
>
> — Thomas and Cook (2005)

Two key challenges for interactive visualization are data size and data complexity. The problem of data size relates to the typically large number of data items to be visualized. In technical terms, we have the possibility to address limited memory, computing power, and display resources with cloud storage, parallel-processing clusters, and large high-resolution displays. However, there is still human perception and cognition, which is not easily changed, if at all. While human visual processing is largely parallel, conscious dialog is possible only with a fraction of the available information. Moreover, complex data contain many different facets and as complexity increases there are more and more questions that one might ask about the data. Any attempt to create a single visual representation that contains all data items, covers all data facets, and provides answers to all questions is condemned to fail. Such a visual representation would be hopelessly overloaded, confusing, and could hardly be interpreted.

Instead, the big problem has to be split into smaller pieces. In a kind of divide-and-conquer way, the pieces become more effective because they are tailored to emphasize a particular aspect of the data, allowing users to concentrate on task-relevant questions. Interaction is the glue that holds the pieces together. Interaction allows users to navigate between different pieces and mentally combine them to form a greater picture. An iterative process takes place during which different parts of the data are brought to the display with emphasis on different facets of the data to provide answers to different analytic questions.

The exploration process generally follows the *visual information seeking mantra* by Shneiderman (1996): "Overview first, zoom and filter, then details on demand." According to this mantra, a visualization should start from an *overview* of the data.[1] From there, the user *zooms* into interesting parts and *filters* out irrelevant items. When necessary, it is possible to descend into the *details on demand*. The exploration can now be continued in different ways. Moving on to data that are related or similar can be a sensible option. Alternatively, one could also return to the overview and investigate the data from a different point of view or with regard to a different question. Taken together, the user forms a mental model of the data by interactively moving from

---

[1]Starting with details and aggregating them to overviews is possible as well (van den Elzen and van Wijk, 2014).

one focus to the next, where the term focus includes data subsets, data facets, analysis questions, and so forth. Eventually, the mental model will be the basis for a comprehensive understanding of the underlying data.

LEVELS AND MODELS OF INTERACTION

Shneiderman's mantra describes the general procedure for visual and interactive information seeking. This procedure involves interaction at different levels. Ware (2012) explains interaction in visualization based on interaction loops at three levels. At the lowest level is the *data manipulation loop*. At this level, interaction is concerned with basic operations of recognizing, pointing at, or manipulating objects (e.g., moving the mouse and performing a click to pinpoint a data item of interest). At an intermediate level there is the *exploration and navigation loop* in which the user combines basic low-level operations to activities of exploration and navigation of large visual data spaces (e.g., adjusting the visual encoding or visiting different parts of the data). At the highest level, the *problem-solving loop* captures processes of forming, refining, and falsifying hypotheses about the data, which involves combining several activities to accomplish cognitively more demanding tasks (e.g., laying out pieces of information for finding relations).

Low-level Interaction   From a conceptual point of view, different modes of interaction can be identified depending on how low-level operations are performed. Spence (2007) distinguishes *stepped interaction* and *continuous interaction* (and passive interaction). Stepped interaction is related to discrete, infrequent actions. Examples are clicking a button or pressing a key to trigger the execution of a clustering algorithm.

Continuous interaction, on the other hand, denotes interaction for which the interaction-feedback loop is iterated at high frequency, continuously so to say. Continuous interaction is vitally important in visualization scenarios as it facilitates examining the visualized data with respect to multiple *what-if* scenarios in a short period of time. Apparently, continuous interaction requires easy-to-execute interactions and sufficient visual feedback, which must be provided quickly. A classic example of continuous interaction is to drag sliders to adjust *dynamic queries*, while the visualization is continuously updated (Ahlberg et al., 1992; Shneiderman, 1994).

Elmqvist et al. (2011) expand on continuous interaction and propose the concept of *fluid interaction*, which integrates aspects of promoting flow, supporting direct manipulation, and minimizing the gulfs of execution and evaluation. Heer and Shneiderman (2012) underline the significance of the continuous and direct character of interaction in visualization: "To be most effective, visual analytics tools must support the fluent and flexible use of visualizations at rates resonant with the pace of human thought."

Intermediate-level Interaction   In addition to thinking of interaction as stepped or continuous low-level operations, it makes sense to consider the intermediate-level exploration and navigation activities typically required to accomplish visualization tasks.

From an analysis of existing interaction techniques, Yi et al. (2007) condense seven categories of user intents for interaction: *select* – mark something as interesting, *explore* – show me something else, *reconfigure* – show me a different arrangement, *encode* – show me a different representation, *abstract/elaborate* – show me more or less detail, *filter* – show me something conditionally, and finally *connect* – show me related items. For each intent, several interaction techniques are listed that support the intent in one way or the other.

Ward et al. (2010) discuss interaction in the context of a framework of interaction spaces (Ward and Yang, 2004). They formulate *interaction operators* (e.g., navigation operators, selection operators, or filtering operators) in analogy to the objectives of an interaction. Further, *interaction operands* are defined as the spaces in which the interaction takes place (e.g., screen space, data value space, or attribute space). Interaction techniques can then be explained by combining operators and operands.

Sedig and Parsons (2013) approach exploration and navigation activities from a pattern-oriented perspective. They suggest looking at action patterns, of which they describe 28 exemplars, including *unipolar actions* such as selecting, navigating, searching, or comparing, as well as *bipolar actions* such as collapsing/expanding, composing/decomposing, inserting/removing, or animating/freezing. Examples are given to illustrate how different interaction techniques instantiate the various patterns.

Spence (2007), Ward et al. (2010), and Ware (2012) elaborate on actual techniques, classic and contemporary ones, for interacting with visual representations and the data behind them. The examples given in these books document the versatility of the existing approaches. The concrete set of techniques to be made available to the users of a visualization solution mainly depends on what the users are expected to accomplish via interaction.

High-level Interaction Similar to how low-level operations are combined to carry out intermediate-level activities, so are the intermediate-level activities a precursor to high-level problem solving. At this level, interaction is considered more broadly as a catalyst for analytic thinking and discovery, which typically focus on the formation, refinement, and falsification of hypotheses by setting data into relation, comparing pieces of information, or extracting high-level features.

The model by Pirolli and Card (2005) describes the sensemaking process as two loops: the foraging loop and the sensemaking loop. The *foraging loop* is concerned with interactively gathering and extracting information. The subsequent *sensemaking loop* describes the development of mental models through schematization as well as hypothesis generation and validation. Liu et al. (2008) take a closer look at the distributed cognitive processes being active when humans engage in a sensemaking dialog with visually represented information. They suggest that "[...] cognition is more an emergent property of interactions between an individual and the environment through perception and action rather than a property bounded inside an individual." In a related context, Liu and Stasko (2010) describe interaction as a tool for giving meaning to what is perceived, for collecting relevant information, and extracting and storing interesting findings.

So, at the highest level, interaction is more concerned with the dialog of the analyst and the knowledge artifacts extracted via interactive visual methods. This also involves interactively coordinating and organizing pieces of information in an analytic visual-interactive workspace.

Structuring interaction according to different levels certainly helps in thinking about interaction. Yet, we should also note that the borders between the levels are not always crisp and clear. Where intermediate-level exploration and navigation ends and where high-level problem-solving loop begins typically depends on the concrete visualization setting. Therefore, it is better to think of smooth transitions between the different levels.

BASIC INTERACTION TECHNIQUES

Interaction techniques are the basic building blocks to complement the visualization pipeline with respect to interaction. An interaction technique typically supports a well-defined intermediate-level activity (e.g., scrolling). To provide its service, an interaction technique relies on several low-level operations (e.g., mouse pointing or multi-touch gestures). Allowing multiple interaction techniques to be used in concert facilitates higher-level interaction (e.g., comparing and annotating findings).

Visualization can draw on a variety of interaction techniques. It is beyond the scope of this section to review them all. Instead, for the purpose of illustration, we describe three classic techniques that are particularly useful to cope with data size and data complexity: Zoomable user interfaces, selection (aka. brushing), as well as dynamic queries.

Zoomable user interfaces   The visual information–seeking mantra suggests that it is necessary to look at data at different levels of detail. In what has become known as *zoomable user interfaces* (Bederson and Hollan, 1994), the user can freely explore the information space at multiple scales. The conceptual idea is that the display serves as a window into the information space (Furnas and Bederson, 1995). Different parts of the information space can be accessed by moving the window, whereas resizing the window adjusts the scale of what can be seen. Bederson (2011) reports that there are various ways of carrying out these operations. Using mouse wheel or mouse drag are common on desktop computers. Pinch and drag gestures are the quasi-standard on touch-enabled devices.

Typically, one differentiates *geometric zooming* and *semantic zooming*. For geometric zooming, scale adjustments will only affect the size of objects on the display. Semantic zooming goes beyond resizing and allows for any kind of adjustments of the visual representation of objects. This makes it possible to show the same data entirely differently at different scales. Yet, care has to be taken to maintain the expressiveness and effectiveness of the visual representation.

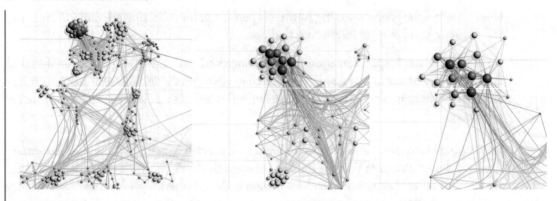

Figure 2.4: Zooming in onto a cluster in a node-link representation of a graph.

A simple example is shown in Figure 2.4. The figure contains a sequence of zooming operations toward a cluster (shown in red) in the upper left part of a graph. The zooming appears to be geometric. On second sight, it is not! In fact, the zooming takes effect on the positions of nodes, but deliberately not on their size. This is important for visualization applications that use size as a visual variable to encode data values (here size corresponds to node degree). Implementing the zoom as illustrated enables the user to see details in dense regions while the integrity of the visual encoding is maintained during zoom operations.

In order to keep users oriented while zooming, visual navigation cues can be provided. Figure 2.4 illustrates the use of a grid for indicating the current scale. Overview windows are particularly helpful for orientation (Hornbæk et al., 2002) and animation can make zoom operations comprehensible (van Wijk and Nuij, 2004). Scroll bars for indicating the current position in the information space and off-screen techniques for hinting at objects currently not being visible are useful as well, not only as visual cues, but also as additional controls of the user interface.

Selection (aka. Brushing)   With zoomable user interfaces it is possible to explore an information space. Selection enables users to further focus their exploration and analysis by pinpointing items of interest in the information space.

As a starting point, we consider the node-link representation in Figure 2.5(a). Our goal is to focus on the connectivity of the high-degree nodes shown in orange and red. Single-item selection is usually carried out by positioning a pointer on top of an item and pressing a trigger. When multiple items are to be selected, drag operations can be employed to define a *rubber band* selecting all items within an elastic rectangle (see Figure 2.5(b)). Greater control is offered by free-form *lasso* selection. Yet, the greater control comes at higher interaction cost for defining the free-from selection. While in all these cases the *absolute position* of items determines their inclusion in the selection, there are also approaches that consider the *relative position* of items.

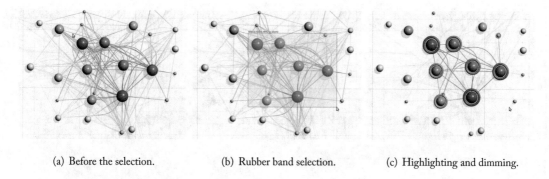

(a) Before the selection.   (b) Rubber band selection.   (c) Highlighting and dimming.

Figure 2.5: Selecting high-degree nodes in a node-link representation of a graph.

An example is *angular brushing* by Hauser et al. (2002) for which selections are determined based on angles between data items.

Once a selection has been performed, the items of interest have to be marked. In Figure 2.5(c), visual highlighting is achieved by placing circles around the selected nodes. Additionally, the selected data is emphasized by dimming the links between unselected nodes. In our example, the visual feedback is shown in the same view where the selection took place. *Brushing & linking* (Becker and Cleveland, 1987; Buja et al., 1991) is a concept to provide feedback in multiple views. To this end, the selection is made a part of the data model and all views being attached to the data model update themselves automatically once the selection has changed. As the data of interest stand out in all views, the linking step provides valuable support for mentally combining different views of the data.

With the selection made in Figure 2.5, we could not fully achieve our goal of focusing on high-degree nodes. Our selection missed some high-degree nodes (in orange), but included some low-degree nodes (in green). So we need mechanisms for altering a selection. Wills (1996) identifies 524,288 theoretically possible ways of combining selections. For the sake of consistency, he recommends that visualization systems should support one of the following five sets of operations: toggle only, replace/toggle, add/subtract, add/intersect, or replace/toggle/add/subtract/intersect. In contrast to that, Chen (2004) proposes complex compound brushing, including logical combinations and analytical filters.

A difficulty with the selections described so far is that they are based on the position of data items on the display. However, the position does not necessarily correspond very well to the aspect according to which we wish to perform the selection (node degree in our case). A promising alternative is to do data-centric dynamic queries.

Dynamic Queries   Dynamic Queries are similar to selection in that the goal is to focus on items of interest (Ahlberg et al., 1992; Shneiderman, 1994). Yet, while selections are defined on the visual arrangement of the data, dynamic queries are specified directly on the data. In the previous

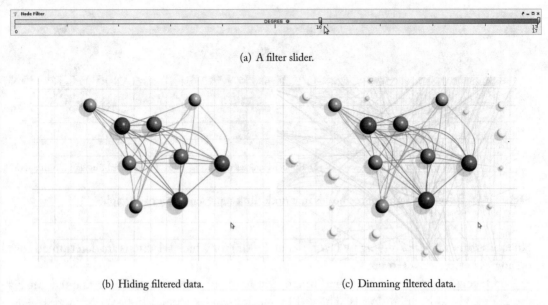

(a) A filter slider.

(b) Hiding filtered data.   (c) Dimming filtered data.

Figure 2.6: Filter slider and visual feedback for dynamic queries.

paragraphs, we used selection to highlight relevant items. Now we use dynamic queries to reduce the number or the visibility of irrelevant items.

Dynamic queries are commonly specified via dedicated filter slides. A filter slider is associated with a data attribute on which the query is to be performed. Handles at the sliders indicate filter thresholds and afford their dynamic adjustment. Figure 2.6(a) shows such a filter slider with two handles for interval queries on the attribute DEGREE. The filter thresholds are set so that low-degree nodes are filtered out. Multiple filter sliders can be combined to create logical AND and OR combinations of filters on multiple data attributes.

A key aspect of dynamic querying is that the visualization is constantly refreshed to provide visual feedback while filter thresholds are being adjusted. There are two options for communicating the filtering result. One option is to hide the data that have been filtered out. This way, irrelevant data is made invisible and the visualization is focused entirely on relevant data as shown in Figure 2.6(b). In some scenarios, it can also be useful to maintain awareness of the fact that there is more in the data than what is shown. In such cases, filtered data can be dimmed to leave residue in the visualization. Figure 2.6(c) shows that dimming helps maintain the overall context.

Discussion   Zoomable user interfaces, selection, and dynamic queries are classic interaction techniques in visualization contexts. Conceptually, they cover basic navigation and selection. Zoomable user interfaces allow the user to *navigate* between different parts of the data and between overview and details. Selection naturally enables the user to *select* data according to their position on the screen. Dynamic queries can be used to *select* data based on data characteristics. In a way, navigation and selection are fundamental, because they are required for many visualization problems. In fact, focusing on relevant parts of the data and interesting data items via zooming, selection, and filtering is often required before more elaborate, intermediate-level or high-level interactive operations can be applied on the focused part of the data.

In scenarios where we deal with well-formed discrete data objects visualized via 2D graphics (e.g., data tables of multivariate data, nodes and edges of a graph layout, regions of maps, words in documents), navigation and selection are relatively easy to implement. Yet, there are also scenarios where no such discrete objects exist, or they are only vaguely defined. This is typically the case for dense visual representations such as plots of scalar functions on maps and image-based depictions of flow data. Going from 2D to 3D graphics (e.g., visualization of volume data) further increases the efforts required for selecting data and defining viewpoints onto the data. Visual representations with vast numbers of small elements, such as point clouds, are also problematic. More sophisticated methods are required to allow selection under such circumstances. Examples include contextual picking (Kohlmann et al., 2009) and what-you-see-is-what-you-get selection for volume visualization (Wiebel et al., 2012), as well as structure-aware selection in 3D point clouds (Yu et al., 2012). A survey on 3D interaction in general is provided by Jankowski and Hachet (2013).

While there is general agreement about the importance of fundamental interactions, there is also recognition that we could benefit more from interaction. Recent research activities strive to take full advantage of interaction as a means to integrate the user more tightly into the visual analysis process. Next we briefly discuss related research questions.

> Interaction in visualization is a special form of human–computer interaction. Interactions are primarily mediated through visual representations of data, which also serve to provide visual feedback.

2.4   THE VISUALIZATION–INTERACTION GAP

Arguably there is still a discrepancy between interaction research and visualization research (Ebert et al., 2014). Zudilova-Seinstra et al. (2009) even see "a major communication gap between HCI experts and those developing visualization algorithms and systems." As a consequence, research results from one field do not always transfer smoothly to the respective other field. Why is this so?

Traditionally, interaction research is focused on the human being. Human-centered design methodologies, models of perception and cognition, accessible devices, as well as social aspects are among the primary contributions of interaction research. On the other hand, visualization research has the focus more on the computer. Automatic extraction of data features, algorithms for the visual mapping, implementations of visualization systems, and high-performance graphics via

GPU acceleration are among the primary topics that can be found in the visualization literature. This is not to say that either side neglects the human or the computer. In fact, visualization is about computer-generated visual representations *for* the human, and in turn, computers are often the enabling technology *for* interaction. But undeniably, the foci are different.

Looking more closely, there is another difference. It regards the use of the visual channel. The visual channel generally has to serve both the graphical interface and the visual representation of the data. Yet, interaction and visualization researchers usually deal differently with the resulting conflict over visual resources. In interaction research, the interface between human and computer is of primary interest. Visual aspects play an important role in terms of the design of the graphical interface. In visualization research, the visual representation of the data is the primary concern, and fewer visual resources are devoted to the graphical interface. Again this is not to say that either side neglects data representation or interface design. In fact, an ideal application would use direct manipulation interfaces that are embedded in the visual representation itself. But again, the foci are different.

Keefe (2010) further contrasts research on interaction in visualization and more general interaction research by considering the specifics of the data, the tasks, and the users. He identifies two key factors that make interaction in visualization different: "(1) complex analysis tasks defined by a specific, highly motivated user population and (2) complex data." Keefe makes his point in a context where experts work intensively with the data. The contrast is less clear in contexts where casual users apply visualization methods (Pousman et al., 2007). But still, differences can be recognized.

Efforts have been made to bring interaction and visualization research closer to each other (Fikkert et al., 2007; Ebert et al., 2014). Keefe (2010) states that the "momentum recently seems to be increasing toward integrating visualization research [...] with interaction research [...]." From a visualization perspective, an ideal world would see visualization tools with expressive and effective visual representations and with interfaces that are usable and enjoyable. Yet, reaching this ideal state requires further intensive research.

RESEARCH AGENDAS

The increasing importance of interaction for visually driven analytical methods has led researchers to start thinking more deeply about the role of interaction. This topic arose in the context of *visual analytics*—the science of analytical reasoning supported by interactive visual interfaces (Thomas and Cook, 2005). In their book, Thomas and Cook describe interaction as "the fuel for analytic discourse" and convincingly explain why developing a new "science of interaction" in the context of visual analytics has top priority. Several research agendas have been proposed by various researchers in the broader scope of interaction in visualization and visual analytics.

Pike et al. (2009) identify research challenges for interaction, covering ubiquitous, embodied interaction, capturing user intentionality, knowledge-based interfaces, collaboration, principles of design and perception, interoperability, as well as interaction evaluation.

Elmqvist et al. (2011) build upon direct manipulation and fluid interaction and envision future interaction research toward the creation of an interaction exemplar repository, the identification of visualization design patterns, and the introduction of visualization criticism.

Lee et al. (2012) point our attention to the mismatch of available interaction technology and the level to which visualization has taken advantage of it so far. They define research challenges related to utilizing modern interaction modalities, providing freedom of expression, taking into account social aspects, breaking down barriers between humans and technology, and gaining a better understanding of human behavior.

Isenberg et al. (2013b) consider data visualization on interactive surfaces specifically. They focus their discussion around technical challenges (surface types and multi-display environments), design challenges (data representation on and interaction with surfaces), and social challenges (collaboration and evaluation).

These studies provide excellent analyses of the status quo of interaction in visualization and enthusiastic calls to action for more research on interaction in visualization. As such, they help strengthen the interaction side of visualization. The level to which interaction can be raised will depend on the concrete responses to the identified research challenges in the form of new interaction concepts, models, and techniques for visualization. A necessary requirement is to narrow the visualization-interaction gap.

2.5   INTERACTION—USEFUL OR HARMFUL?

Much has been done in recent years helping us to better understand how interaction in visualization actually works and how it affects the ability of human beings to extract knowledge from data. The many references cited in the previous sections are generally in favor of interaction and recognize it as a strong positive factor of visualization. But there are also critical voices. We shall discuss this matter very briefly in order to make the reader aware of the potential disadvantages of interaction. This is not to cast a poor light on interaction, but rather to underline the importance of thinking carefully about interaction in visualization.

USEFUL INTERACTION

Let us start positively. The human-in-the-loop argument is often brought forward, putting the human in the position to make the final decisions, rather than leaving it to the computer (Thomas and Cook, 2005). Examples of useful interaction can be found en masse in the literature. Empirical evaluation in the form of quantitative or qualitative studies testify to the benefit of interaction. For example, in a longitudinal study by Saraiya et al. (2006), "methods to efficiently interact [...] were considered equally or even more important than the visual representations." For another example, Amini et al. (2015) found evidence of the positive effect of interaction in 2D and 3D visualization scenarios.

In fact, interaction enables the human to generate or influence results in a way that goes beyond what is computable. Wegner (1997) offers an explanation for why interaction is more powerful than algorithms:

> "Algorithms are metaphorically dumb and blind because they cannot adapt interactively while they compute. They are autistic in performing tasks according to rules rather than through interaction. In contrast, interactive systems are grounded in an external reality both more demanding and richer in behavior than the rule-based world of noninteractive algorithms."
>
> — Wegner (1997)

Wegner's discussion is based on a theoretical model of *Interaction machines*, which he demonstrates to be more powerful than *Turing machines*. Such theoretical considerations further strengthen the already positive picture that is generally drawn of interaction.

HARMFUL INTERACTION

On the other hand, a recent definition of *interactivity* as the *quality of interaction* (Sedig et al., 2012) suggests that there is a spectrum of interaction with positive and not-so-positive instances. Looking at Norman's (2013) gulfs of execution and evaluation in Figure 2.3 back on page 8 offers an explanation. If the gulfs are not bridged or at least sufficiently narrowed, interaction can be a hindrance.[2]

Cooper et al. (2007) identify the mismatch between the *implementation model* and the *user model* as a primary source of interaction problems. To minimize problems, the *represented model*, which serves the interface, should be as close as possible to the user's mental model while abstracting as much as possible from the implementation model. The illustration in Figure 2.7 makes this point clear.

In the context of visualization, the involved models are often complex and matching them is not always easy. Failing in this regard most likely leads to suboptimal or even bad interaction, of which several examples exist in working visualization software prototypes, but which are hardly reported in scientific publications.

In addition to concrete examples of bad interaction, there can be general reservations about interaction. Tominski et al. (2011a) report on the results of a questionnaire that was carried out to assess the distribution and use of interactive visualization tools in climate research. An observation particularly related to interaction was that there can be a kind of mistrust in interaction in general. The participants feared the arbitrariness of visual representations that have been generated by interactive adjustments of thresholds or visualization parameters. The reason was that it is no longer clear if an artifact identified in the picture is an actual feature in the data or just a pattern-by-chance. Such statements urge us to provide interaction in a balanced way, offering flexibility, but within reasonable boundaries to avoid arbitrariness.

---

[2]Interestingly, even hard-to-use interaction can be beneficial under some circumstances (Riche et al., 2010b).

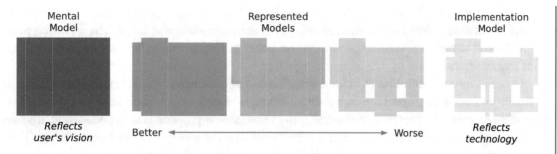

Figure 2.7: The represented model should be close to the user's mental model and abstract from the technological implementation details (adapted from Cooper et al., 2007).

There are even voices that openly challenge interaction in an attempt to break the myth of interaction as a universal cure. Victor (2006) discusses interaction in the context of software to learn, to create, and to communicate, which are activities that are certainly addressed by visualization software as well. While interaction is indispensable for creating (e.g., constructing a CAD model), interaction for learning (e.g., accessing information from a visual representation) "is considered harmful." Victor argues that only as a last resort should input be solicited from the user. Considering the costs involved when interacting in visualization settings (Lam, 2008), this is a sensible argument. Victor's critique mainly relates to the tendency of being quick with answering a particular user need with providing a way to satisfy it interactively. But it is the task of the system (and the designer of the system beforehand) to provide the information needed in a particular situation. This suggests to us think of interaction in a *less-is-more* way, and to infer, where possible, how interaction costs can be reduced.

Still, an overall positive image of interaction in visualization shall be maintained. To this end, we have to strive for useful interaction and avoid harmful interaction. We have to design and implement visualization approaches according to accepted criteria and models of interaction and under consideration of the requirements of visualization. In the context of large data and complex analytic tasks, this is not an easy endeavor. While visual representations of data are usually developed based on explicitly addressing visualization design rules, the rationale behind the interaction techniques to work with the data at times remains vague. To overcome this deficiency it makes sense to frame in more detail the aspects that influence interaction in visualization. In Chapter 3, we identify and discuss these aspects and compile them into an encompassing interaction-oriented view on visualization.

2.6   IMPLEMENTING INTERACTIVE VISUALIZATION

So far, we have considered interaction as a means for the user to steer the visualization and to engage in an analytic dialog with the represented data. We also reflected on useful and harmful interaction, briefly touching the domain of interaction design. Yet, interaction must also be implemented as a workable solution in order to be actually usable and useful. The following paragraphs will therefore review some aspects and issues of implementing interaction in the context of visualization.

The fundamental model behind visualization approaches is the visualization pipeline. It describes how data (input) are transformed through several stages (processing) to generate visual representations (output). Technically, the visualization pipeline is well understood, and accepted software design patterns exist for implementing the pipeline in working software (Heer and Agrawala, 2006; Moreland, 2013). When developing visualization software, one can resort to powerful libraries and toolkits, such as VTK (Kitware, Inc., 2010), Obvious (Fekete et al., 2011), or D^3 (Bostock et al., 2011).

Capturing the technical essence of interaction in visualization, Jankun-Kelly et al. (2007) propose the *p-set model* of visualization exploration. This model abstractly explains user interaction as operations that change visualization parameters. In other words, any concrete interaction condenses down to the adjustment of parameter values. Visualization parameters can be manifold, e.g., the viewing angle into a 3D scene, the focus point of an interactive lens, thresholds of a dynamic query operation, or parameters that control a clustering algorithm. Modeling interactive exploration abstractly as parameter changes is not only a useful basis for implementing interaction, but also for maintaining undo and redo histories (Kreuseler et al., 2004) and for realizing advanced mechanisms for coordinated interaction (Weaver, 2004).

While visualization can be modeled as a parametrized *pipeline*, the implementation model for user interaction is a *loop*. This makes developing interaction for visualization (i.e., action-feedback loop) fundamentally different from developing visualization algorithms (i.e., linear input-to-output processing). The model-view-controller (MVC) pattern (Krasner and Pope, 1988) is widely applied to structure the essential high-level software components. The *model* encapsulates the data and parameters. The *views* visualize the data under consideration of the parameters. The *controller* handles interaction.

The difficulty lies in coordinating the components. Edwards (2009) coins the term *Callback Hell* in the broader context of reactive systems. For interactive visualization systems, the controller logic is typically complex due to the comprehensiveness of the interaction. Multiple callbacks or event handlers have to work together to realize the rich interactive behavior in a coordinated effort. Already for classic mouse and keyboard interaction, implementing and maintaining the event-reaction logic can be difficult. For touch interaction with multiple pointers or multi-modal interaction it gets even worse.

A basic interaction loop is illustrated below. For the sake of simplicity, the usually complex event handling is hidden in line 5 in the function $update(e)$, which updates the underlying model according the event e. Events can be input events (e.g., mouse move, touch down, or key press), but also timer events for animation or other system events.

Algorithm 1 Basic interaction loop.

1: *init*()
2: **repeat**
3:   **if** *hasEvent*() **then**
4:     *e* = *getEvent*()
5:     *update*(*e*)
6:   **end if**
7:   *render*()
8: **until** *e* = END

There are approaches that can facilitate the implementation of interaction. For example, state machines can be used to overcome event handler spaghetti code (Appert and Beaudouin-Lafon, 2008). The model-display-picking-controller (MDPC) pattern—an extension of the MVC pattern—separates display and interaction geometry to ease the specification and the implementation of interaction (Conversy, 2011). For the definition and detection of multi-touch gestures one can use tablatures and regular expressions (Kin et al., 2012). While the mentioned approaches are not specifically for visualization, they are still definitely worthwhile to consider when developing interactive visualization software.

2.7   SUMMARIZING THOUGHTS

The previous sections looked at interaction from different points of view. We covered general aspects of visualization and human–computer interaction, different levels of interaction in visualization, basic interaction techniques for visualization, the visualization-interaction gap, the question of good and not-so-good interaction, as well as the issue of implementing interactive visualization software. For a summarizing discussion of the key points, we refer to Figure 2.8.

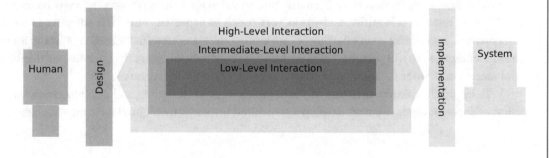

Figure 2.8: Different levels of interaction flanked by interaction design and implementation.

Interaction in visualization has to be approached from a human-centered angle. In Figure 2.8, this is illustrated by a human and a box labeled *design*. The design of useful interaction must consider the human sensory and motor skills as well as the data exploration activities and sensemaking tasks that humans are engaged with.

Interaction in visualization has to be considered from a technology-oriented perspective as well. This perspective is reflected by a system and a box labeled *implementation* in Figure 2.8. Implementing interaction means transferring designs into workable tools that utilize computer technology to manage, analyze, transform, and represent data by interactive visual means.

There is a mutual dependency between designing and implementing interaction. Well-designed interaction is rendered useless with insufficient implementation and the best architectures lie fallow without appropriate interaction designs.

In visualization scenarios, we have to consider multiple levels of interaction as illustrated in the center of Figure 2.8. Addressing low-level operations is fundamental because they have definite impact on the higher levels of interaction. But focusing on lower-level aspects alone is not sufficient. Intermediate-level activities must be taken into account and these must be put in the context of higher-level analytic thinking.

Designing and implementing interaction in visualization at multiple levels is the topic we aim to shed some light on in this work. To give a brief outlook, we will be concerned with the implementation aspect when we introduce a multi-threading visualization architecture addressing the computational requirements for fluid interactive visual exploration. In terms of interaction design, we will consider low-level and intermediate-level interaction, and also touch aspects of high-level interaction. We will look at novel ways of how to fundamentally interact with visualizations by means of tangible views and physical navigation. Methods for intermediate-level exploration and navigation will be introduced in the context of graph visualization and visualization of spatio-temporal movement data. Further, we will discuss interaction support for more high-level visual comparison tasks and semi-automatic editing of graphs with customized layouts.

Across all levels, the goal is to provide "good" interaction that is useful, usable, and also enjoyable. Being aware of the fundamental models of interaction as well as the potential down sides of interaction is a necessary requirement. We will illustrate how "good" interaction can be fostered by designing according to humans' natural behavior and by integrating assistive methods. "The interaction is intuitive and works very much as expected" and "The software is nicely implemented, everything is very harmonic" are among the statements users made about the interaction techniques we will present. Such enthusiastic feedback suggests that our line of thinking can indeed lead to useful, usable, and enjoyable interaction for visualization.

To establish a clearer structure of the complex topic of designing and implementing interaction at different levels, we next develop a structured interaction-oriented view on visualization.

CHAPTER 3

Aspects of Interaction in Visualization

The goal of this chapter is to define a view on visualization that is focused on interaction. This interaction-oriented view shall be organized around the following key aspects:

- Data

- Tasks

- Technology

- Human

- Implementation

The *data* are a primary concern of visualization and so they are a key factor of interaction as well. The second aspect are the *tasks* that have to be supported by visualization tools. Data and tasks together can be considered the core of our view as illustrated in Figure 3.1. This core is flanked by the *human* and the *technology*. The human and the technology carry out the given tasks on the data in a cooperative effort. The human is recipient of visual representations generated by the technology. The technology provides interaction modalities allowing the human

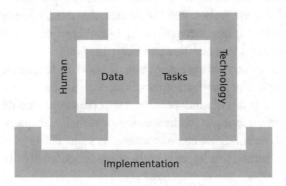

Figure 3.1: The data and the tasks are core concerns flanked by the human and the technology. Together they are grounded in the implementation.

to act upon what is perceived from the visual representations. Sensemaking is usually left to the human, whereas complex computations are taken care of by the technology. Finally, an efficient *implementation* is needed for a smooth interplay of human and computer. The implementation transfers the visual and interaction design into an operational software that can actually be used to carry out visualization tasks.

Taken together, data, tasks, technology, and human can be understood as cornerstones of interaction grounded in an efficient implementation. In the following sections, we discuss the individual aspects in more detail.

3.1   THE DATA

Data characteristics have long since been identified as a factor that influences the visualization design. From a practical perspective, one can differentiate several *data types*: temporal data, geospatial data, volume data, flow data, multidimensional multivariate data, graph data, and text and document data. The differentiation of data types naturally led to different classes of visualization techniques and sub-communities in the visualization realm (e.g., flow visualization, volume visualization, graph visualization, geographic visualization, or time-series visualization).

From a more mathematical point of view, data can be differentiated based on *data scale* into qualitative and quantitative data. Qualitative data can further be divided into nominal and ordinal data. For nominal data, it is only possible to determine equality of data values. For ordinal data, an order is defined allowing us to determine if a value is greater or less than another value. In contrast to qualitative data, for quantitative data, we can additionally quantify the difference between values, that is, we can determine the distance between values. Quantitative data can be discrete (i.e., integer-valued) or continuous (i.e., real-valued).

When transforming data into visual representations, it is general practice to use visual variables (Bertin, 1983) that suit the data scale. Mackinlay (1986) recommends good matches of data scales and visual variables based on empirical studies by Cleveland and McGill (1986). For example, position and size are well-suited for quantitative data, whereas shape and hue play increasingly important roles for qualitative data. The validity of these recommendations has been confirmed more recently by Heer and Bostock (2010).

When we look at the interaction side of visualization, unfortunately no such accepted datacentric view can be found. For low-level interaction, such a data-centric view might not be necessary, because this level is usually independent from data type and data scale. But already for the fundamental dynamic queries, it makes a difference whether a query slider operates on a data attribute with nominal, ordinal, discrete, or continuous scale. Riche et al. (2010a) investigate this issue in the context of interactive legends with query functionality. They consider data scale a relevant design factor.

In terms of data types, there are obvious differences as well. For example, interactively navigating temporal data along the dimension of time is different from navigating along movement trajectories in geographical space. For another example, selecting a subset of flow vectors (e.g., vectors with specific direction and magnitude) is different from selecting a subset of graph nodes (e.g., the k-neighborhood of a node of interest).

We do not yet have a full understanding of the implications of data characteristics on interaction in visualization. In the visualization literature, not much is reported on how interaction can be attuned to the data. But there are examples that illustrate how data-aware interaction design can support users. Techniques for interacting with visual representations of time-oriented data demonstrate this quite well (Hochheiser and Shneiderman, 2004; Holz and Feiner, 2009; Zhao et al., 2011). Yet, more research is needed, not only for other types of data, but also for dealing with the complexity and multifaceted character of today's datasets.

> The interaction design for visualization should consider the character of the *data*, very much as the design of the visual encoding already does.

3.2   THE TASKS

In addition to considering data characteristics, it is also important to address the tasks to be accomplished with interactive visualization. A classic high-level differentiation is to distinguish *exploration* for forming hypotheses, *confirmation* for falsifying hypotheses, and *presentation* for communicating findings (Ward et al., 2010). More concrete lists of tasks are compiled by Shneiderman (1996), Zhou and Feiner (1998), and Amar et al. (2005). Their lists include tasks such as retrieve value, filter, relate, compare, rank, find extremum, cluster, and correlate, to name only a few. Tasks can also be specific to a particular type of data. Andrienko and Andrienko (2006) present a fine-grained and formal analysis of tasks related to spatio-temporal data. Lee et al. (2006) look at the specifics of tasks being relevant when visually analyzing graphs. A comprehensive design space of visualization tasks is described by Schulz et al. (2013a).

Tasks are generally accepted as an influencing factor for the visualization design. There are a number of visualization examples that have been specifically designed according to given tasks. Treinish (1999) discusses the task-specific design of visual representations for weather forecast. Tominski et al. (2008) use different color scales for identification, localization, and comparison tasks. The benefit of such task-oriented approaches is that the visual representation suits the task at hand.

On the interaction side of visualization, there are a few studies that investigate the task aspect. Early work by Chuah and Roth (1996) considers the semantics of basic and composite tasks in interactive visualizations. We already mentioned the work on user intents for interaction by Yi et al. (2007) and the rich set of patterns of interaction for complex cognitive activities with visual representations by Sedig and Parsons (2013). Again, tasks can be data-specific as with the taxonomy of cartographic interaction primitives by Roth (2013) or the interaction task for multivariate graphs described by Wybrow et al. (2014).

These studies show us the principal objectives of interaction. Moreover, mappings can be established between the objectives and interaction techniques that support them. For example, Yi et al. (2007) naturally associate their *filter* intent with dynamic queries. As an example for their *navigating* pattern, Sedig and Parsons (2013) refer to the interaction techniques for navigating graphs as described later in Section 4.2.1. Such suggestions of concrete techniques allow practitioners to find solutions to given problems easily.

Where we still lack understanding is how to transform our knowledge about interaction tasks into task-driven interaction designs for visualization. There are many examples of interactive solutions that superbly support users in carrying out visualization tasks. However, the design process behind them, including tasks, design alternatives, and design decisions, often remains undisclosed in visualization papers.

Fostering a task-centric view will help us to better attune interaction in visualization to what the users actually aim to accomplish. Yet more research is needed to cover the task aspect comprehensively. A particularly interesting topic is to combine interaction for data *exploration* tasks and interaction for data *editing* tasks (Baudel, 2006; Kandel et al., 2011).

> Analytic exploration and editing *tasks* are a determining factor for the design of interaction as well as for the design of the visualization.

3.3   THE TECHNOLOGY

When we look at the settings in which visualization tasks are primarily carried out these days, we will most certainly see the classic setup where a user is sitting at an off-the-shelf desktop computer. A regular display shows visual representations, while mouse and keyboard are used for interacting with the system. This setup has been predominant for years. Yet, new technologies have emerged in terms of both display devices and interaction modalities.

Modern display devices, such as large high-resolution displays or small hand-held devices, have been addressed by adapting existing visualization approaches or devising new ones. Examples are the works on visualization on small, mobile devices by Hakala et al. (2005), Chittaro (2006), and Ying et al. (2012). Visualization in environments with large and multiple displays is studied by Yost and North (2006), Forlines and Lilien (2008), and Radloff et al. (2015). Their primary goal is to develop visual representations that fit the properties of the displays.

New display environments usually also require alternative ways of interaction. For example, using a mouse to move a cursor across a giga-pixel display is certainly not feasible. Modern interaction technologies, such as body tracking, gaze detection, or sensors for multi-touch and tangibles, offer ample opportunities for rethinking existing visualization solutions with regard to interaction. For example, touch tables and tablets blend display and interaction into a single interactive-visual medium that enables us to make Shneiderman's (1983) direct manipulation truly *direct*. Forming mental models based on interactively exploring and manipulating data directly under one's fingertips on touch-enabled displays appears to be quite a promising prospect.

There is already research exploring the requirements and advantages of touch interaction for visualization (Voida et al., 2009; Kosara, 2011; Klein et al., 2012; Yu et al., 2012; Coffey et al., 2012; Sadana and Stasko, 2014). Other forms of interaction (e.g., sketching) have also been found to be beneficial (Keefe et al., 2008; Keefe and Isenberg, 2013).

Yet, novel interaction modalities only slowly find their way into the visualization mainstream, although they considerably broaden the spectrum of what is possible. There is still a gap in terms of promising new technologies on the one hand, but only little integration of these technologies into visualization research and applications on the other hand (Tominski et al., 2011b; Lee et al., 2012; Isenberg et al., 2013b).

A reason might be that new ways of interacting are not straight-forward to integrate into working models, frameworks, and solutions. Only little do we know about the actual benefits and potential shortcomings of new technologies when coupling them with visualization. Finding appropriate combinations of display technology and interaction modalities and seamlessly integrating the technology into visualization workflows are key concerns. Moreover, there are technical challenges to be addressed. Developing interaction for classic devices is already a complex matter. Taking new and possibly multiple technologies into account further increases the demands in terms of both designing interactions and actually implementing them.

> Interaction in visualization can take advantage of *technology* by integrating modern display devices and new interaction modalities.

3.4 THE HUMAN

From its early beginnings, visualization research has considered effectiveness vitally important. The effectiveness criterion demands that visual representations be designed to conform with the characteristics of the human visual system (Mackinlay, 1986). A deeper discussion of human factors in visualization is provided by Tory and Möller (2004). Contemporary research continues to investigate human aspects as documented by Kerren et al. (2007) and Huang (2013) in the context of human-centered or human-centric visualization.

For human–computer interaction in general, usability (Nielsen, 1993) and user experience (Hassenzahl and Tractinsky, 2006) as described in Section 2.2 are key factors to be considered. In this regard, it is useful to think of Norman's (2013) seven stages of action—forming the gulfs of execution and evaluation—and the costs associated with them. Lam (2008) provides a comprehensive view on interaction costs in the context of visualization. Her framework identifies costs to form goals, to form system operations, to form physical sequences, to execute sequences, to perceive state, to interpret perception, and to evaluate perception. When developing interactive visualization solutions, we have to be aware of these costs and try to minimize them.

Elmqvist et al. (2011) specify several requirements for interaction in visualization. These requirements demand that interaction be fluid in terms of the performance of actions, the presentation of smoothly animated visual feedback, and in terms of switching seamlessly between different tasks as they occur in visualization scenarios. Keefe and Isenberg (2013) go one step further and argue for natural user interfaces in visualization. In doing so, they pick up a hot trend from HCI research in natural interaction (Valli, 2008; Jacob et al., 2008; Wigdor and Wixon, 2011). Although the term *natural* is still debated among researchers, the objective is generally agreed upon: Make interaction with the computer more akin to how humans interact naturally in the real world.

Reducing costs, promoting fluidity, and striving for naturalness are all related to the overall goal of making the cycle of performing interaction and interpreting the system's response smooth and efficient for the human. The interaction must be easy to perform and the visual feedback must be easy to interpret, even in light of complex analytical tasks with feature-rich data visualizations in the background.

> Paying attention to interaction costs, fluidity, and naturalness is important when addressing *human* aspects of interaction in visualization.

3.5   THE IMPLEMENTATION

Algorithmic efficiency has always been an important topic in visualization. Once a new visualization method has been invented, visualization researchers typically seek faster implementations to be able to show larger data at higher frame rates. Offloading extensive computation to the graphics hardware certainly is a common method of improving visualization performance. Examples of successful searches for better algorithms are the FastLIC algorithm for flow visualization by Stalling and Hege (1995) and the improved algorithm for generating Voronoi treemaps by Nocaj and Brandes (2012).

Implementation efficiency is also a critical concern for interaction. When a user interacts with a visualization, rich and informative visual feedback must be presented quickly. This is particularly important for the many atomic changes that happen during *continuous interaction*. The literature suggests a response time below 50–100 ms (Card et al., 1991; Shneiderman, 1994; Spence, 2007). However, depending on data size and data complexity as well as on the influence of visualization parameters, computing rich feedback in a fraction of a second can be challenging. It might be necessary to re-compute the entire visualization pipeline, including analytical processing, visual abstraction, and rendering of potentially many data items and graphical primitives. The risk for interaction is that visual feedback lags behind, disrupting the interaction flow (Liu and Heer, 2014).

Another aspect adds to the time costs for presenting visual feedback. As interaction involves change, we have to take care that users understand what is happening. Abrupt changes in the visual display can be useful to draw the user's attention, but they also may hurt the mental model that users are developing during data exploration. There is evidence that interpolating the parameter change and applying animation to present the visual feedback is a sensible solution (Heer and Robertson, 2007; Pulo, 2007). However, animations take time as well, and the potentially costly interpolation of animation parameters cannot be neglected.

As can be seen from the previous discussion, interaction developers face a two-sided conflict. On the one hand, interaction needs *synchrony*, which comes down to all-time responsiveness of the visualization system and immediacy of the visual feedback. A system that is unresponsive and blocked while computing is a worst-case scenario for the user. On the other hand, interaction needs *asynchrony* for the computation of the visual feedback and for its animation. The difficulty is to integrate synchrony and asynchrony.

These implementation concerns are only a few among many that need to be addressed when developing interactive software in general (Olsen, 2009) and visualization systems in particular (Tang et al., 2004). Therefore, it is relevant to support the human not only in its role as a visualization user, but also in its role as a visualization developer. As a basis for such support, Keim et al. (2010) and Fekete (2013) argue for investigating suitable infrastructures that help in crafting efficient visualization tools. Letondal et al. (2010) and Conversy (2013) provide a more detailed discussion and compile a list of requirements for interaction-oriented development tools. Although their analysis is not specific to visualization, it can be taken for granted that the points raised hold true for the development of interaction in visualization as well.

> Efficiency is crucial for interaction in visualization. Sophisticated *implementations* are needed for effective interaction with the data.

We have now discussed all key aspects that make up our interaction-oriented view of visualization: the data, the tasks, the technology, the human, and the implementation. In the next chapter, we will pick up these aspects and present several approaches that particularly address these aspects.

CHAPTER 4

Methods and Techniques for Interactive Visualization

Structured according to the key aspects discussed in the previous chapter, we will now introduce methods and techniques for interacting with visual representations of data. For each section of this chapter, the focus will be set on a specific aspect. In addition to focusing on individual aspects, we will also establish connections across them, illustrating the overarching character of the addressed problems and solutions.

The beneficiaries of these solutions are mostly the human users. They are the ones who have to work on their tasks related to their data using the technology available to them, and hopefully their work will be effective and effortless thanks to good interaction designs and appropriate implementations.

In another role, humans also act as visualization developers. Addressing their needs and providing supportive approaches and infrastructures is a challenge in its own right. In light of this challenge, we aim to ease the developer's life by presenting a multi-threading visualization

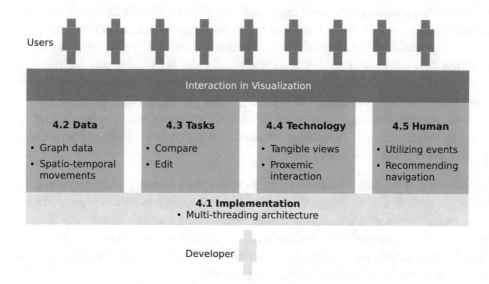

Figure 4.1: Overview of interaction methods and techniques presented in this chapter.

architecture. This architecture can serve as a foundation for interactive visualization solutions, from which, in turn, end-users can profit as well.

Figure 4.1 provides an overview of the aspects of interaction in visualization and the associated withpics we will discuss. The following sections are organized so as to (1) introduce the problem, (2) present the approach, (3) discuss the result, and (4) suggest further readings. To be able to cover all aspects, we have to keep our explanations compact. For further details, we refer the reader to the corresponding original literature. In the next section, we start with addressing the implementation aspect as the basis for efficient interaction in visualization.

4.1   AN ARCHITECTURE FOR EFFICIENT INTERACTIVE VISUALIZATION

The basis for any useful interactive visualization is an architecture that implements the interaction-feedback loop efficiently. Heer and Shneiderman (2012) express the following concerns regarding the engineering of efficient visualization infrastructures:

> "Especially for large datasets, supporting real-time interactivity requires careful attention to system design and poses important research challenges ranging from low-latency architectures to intelligent sampling and aggregation methods."
>
> — Heer and Shneiderman (2012)

Bringing in line *synchrony* (related to responsiveness and immediacy of the visual feedback) with *asynchrony* (related to the time cost involved to generate and present the visual feedback) has already been identified as a major challenge. A straightforward implementation of the classic visualization pipeline will not be helpful in this regard. Any computation along the pipeline that fails to deliver results within interactive response time (ca. 50–100 ms for continuous interaction) will disrupt the interaction-feedback loop, and thus hinder fluid interactive analytic work. What is needed is an architecture that can cope with complex, time-consuming computation and is able to react to interactive user requests at any time, while providing visual feedback as rich as possible. Utilizing the advantages of modern multi-core processors would be one option. However, developing interactive multi-threading solutions is notoriously difficult and prone to manifold implementation errors (Lee, 2006).

Approach   To avoid these difficulties and to make implementing multi-threading less error-prone, Piringer et al. (2009) designed a general multi-threading architecture around the concepts of *early thread termination* and *layered visualization*. Using multiple computing threads accommodates the need for asynchrony, and early thread termination accommodates the need for synchrony. Using multiple visualization layers makes it possible to scale the richness of the visual feedback according to the available computing resources.

The conceptual model of the general multi-threading architecture is illustrated in Figure 4.2. The architecture consists of four principal components: two storage components (the

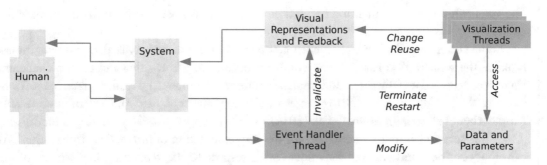

Figure 4.2: Architecture of multi-threaded visualization.

input storage and the output storage) plus two processing components (the event handler thread and the visualization threads).

The input storage consists of the data to be visualized and the visualization parameters, which is analog to the *p-set model* (Jankun-Kelly et al., 2007). The output storage holds the visual representations. These include the visual feedback and the data visualization. The visual representations are not necessary complete, but can be partial if a computation had to be terminated early due to user interaction.

In terms of the processing components, the architecture is based on separation of concerns to be able to cope with synchrony and asynchrony. The key to responsiveness is the dedicated event handler thread. The only responsibilities of this thread are to receive interaction input events from the user and to perform signaling operations with respect to the other components of the architecture. The visualization threads are responsible for transforming input data and parameters into visual output. The visual output is subdivided into multiple layers according to different strategies (e.g., semantic layers, incremental layers, or level-of-detail layers). Using multiple layers enables the architecture to provide rich and scalable visual feedback, and to avoid redundant computations by reusing cached results that remain valid after a user interaction.

Results and Discussion   Empirical studies confirm that the architecture offers significant performance improvements, and hence provides a good basis for implementing interactive visualization solutions. The clear structure of the architecture further helps in avoiding typical programming errors that occur when multiple threads are involved. Successful installments of the architecture (Tominski et al., 2004; Doleisch, 2007; Piringer et al., 2008; Tominski et al., 2009) testify to its usefulness and general applicability across programming language boundaries.

The presented architecture facilitates the engineering side of interaction in visualization and its efficiency affects interaction positively on all levels. More details on the architecture in general and the early thread termination and layered visualization in particular along with a discussion of design choices and empirical studies are provided by Piringer et al. (2009). Actual interaction techniques built on top of a system that instantiates the architecture are described in Section 4.2.1.

The visualization tools we mention in Section 4.5.1 have also been built on the multi-threaded approach.

The currency of the topic of efficient architectures for interactive visualization remains unbroken. The visualization research community continues to address this and related challenges via new concepts, data structures, and implementations. The interested reader is referred to Zinsmaier et al. (2012); Liu et al. (2013); Im et al. (2013); Lins et al. (2013); Cottam et al. (2014); Mühlbacher et al. (2014); Stolper et al. (2014).

This section laid out a foundation for the implementation of interactive visualization. We can now consider methods and techniques with respect to the remaining key aspects of our interaction-oriented view: the data, the tasks, the technology, and the human.

Related Readings
Law et al. (1999) • Weaver and Livny (2000) • Fekete and Plaisant (2002) • Chan et al. (2008) • Moreland (2013)

4.2   DATA CHARACTERISTICS AND INTERACTION

In this section, we focus on the data aspect, that is, on the interplay of data characteristics and interaction in visualization. There are a number of standard interaction techniques that work well without particularly considering the data characteristics. An example is graphical selection (aka. brushing) as mentioned in Section 2.3. Selection is a relatively low-level interaction. At intermediate and higher levels of interaction, it becomes more important to take the characteristics of the data into account.

Here we assume a data model similar to the ones proposed by Kreuseler and Schumann (2002) and Ware (2012). Data are defined as data entities that are associated with quantitative or qualitative data values, and the data are characterized by:

1. the data's *structure* and

2. the data's *frame of reference*.

The structural aspect captures relations among data entities. Structures can be described as graphs in a most general sense. The data's frame of reference captures the spatial and temporal context in which the data have been collected or generated.

We present interaction techniques that particularly consider the structure of the data and the spatial and temporal dependencies of data. Our focus will be on navigation techniques and interactive lenses for exploring graphs and spatio-temporal movement trajectories. The navigation and lens techniques are designed according to the data characteristics and engineered so as to exploit the data characteristics. As such, the presented approaches illustrate quite well the benefit of considering the data aspect a key influencing factor for interaction in visualization.

4.2.1   INTERACTING WITH GRAPHS

Graphs have gained increased importance in many fields of study. A graph is a universal model that helps in structuring and relating entities of interest (Rodriguez and Neubauer, 2010), be it enzymes in biomedical networks, people and their social behavior, or simply pieces of information. The increased importance of graphs led to an increased demand for interactive visualization tools for graph data.

Classic graph drawing is concerned with generating layouts of graphs according to different aesthetic criteria (Battista et al., 1999; Tamassia, 2013). Aspects of interaction with graphs and their visual representations are becoming increasingly relevant. McGuffin and Jurisica (2009) state that

> "[...], there remains a significant need for users to be able to interactively manipulate such graphs, [...]"
>
> — McGuffin and Jurisica (2009)

As documented in surveys by von Landesberger et al. (2011) and Wybrow et al. (2014), graph visualization methods have begun incorporating more advanced interaction techniques. Here we focus on supporting the navigation and exploration of visual representations of complex graphs.

Approach   Addressing node-link layouts in particular, Tominski et al. (2009) develop several interaction techniques for graphs. What they call *radar view* and *edge-based traveling* are interactive visual tools for easy navigation with a preview of what can be expected to be found in the direction of traveling. Novel interactive lenses are presented as tools for graph exploration. The *local-edge lens* and the *layout lens* can be used to tidy up edge clutter and to create local overviews of the neighborhood of nodes of interest on demand.

Techniques for Navigating Graphs   To facilitate the exploration of large graph layouts at different levels of detail, it makes sense to embed the visualization in a zoomable space (see zoomable user interfaces in Section 2.3). In order to fully take advantage of zoomable spaces, users must be supported in finding answers to a number of questions, including, "Where can I usefully go?" and "What lies beyond?" (Spence, 2007). The *radar view* provides answers to these questions.

As illustrated in Figure 4.3, the radar view is a technique that provides a look-ahead in the direction of navigation. All off-screen nodes that fall into the radar are projected onto the view border to make them visible to the user. The radar automatically follows any change of direction initiated by the user with the navigation wheel (bottom-right in Figures 4.3 (a) and (b)). This way, the user can quickly acquire an overview and navigate to potentially interesting candidates to be visited next.

Yet, navigating in larger layouts can be cumbersome and time consuming. This is where *edge-based traveling* comes into play. The key idea is to utilize the structure of the graph as

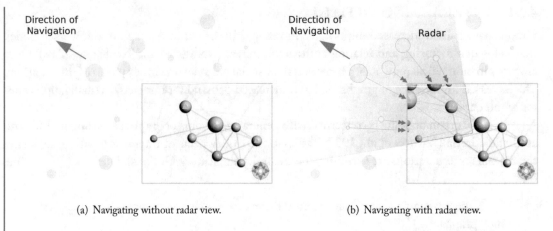

(a) Navigating without radar view.   (b) Navigating with radar view.

Figure 4.3: The radar view projects off-screen nodes to the view border to make them visible.

a kind of railway system across which the user can travel easily from one station to another. Simply clicking on edges triggers navigation to incident nodes. This way, the user can cover great distances without manual zoom and pan operations. Smooth animation makes the navigation comprehensible (van Wijk and Nuij, 2004).

Lens Techniques for Graph Exploration Even sophisticated algorithms cannot guarantee that layouts of larger graphs are clutter-free and that adjacent nodes lie close together. So, users exploring a graph may encounter edge clutter and investigating a node's connectivity may be complicated due to widely distributed neighbors. The interactive graph lenses presented in Figures 4.4 and 4.5 enable users to overcome these difficulties.

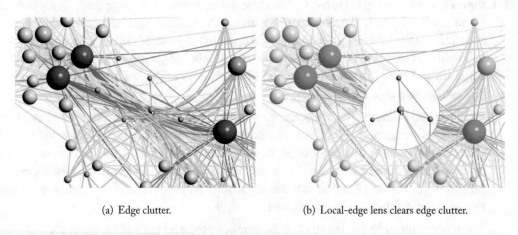

(a) Edge clutter.   (b) Local-edge lens clears edge clutter.

Figure 4.4: The local-edge lens clears the view of edge clutter.

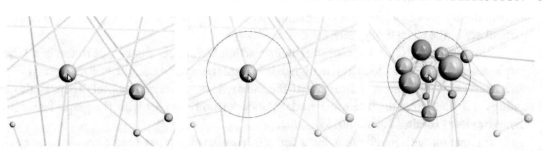

(a) No lens: Connectivity remains un-clear.

(b) Local-edge lens: Edges are visible, but neighbors are not.

(c) Layout lens: Neighborhood visible at a glance.

Figure 4.5: The layout lens brings otherwise invisible neighbors to the lens.

The *local-edge lens* is designed so as to tidy up the lens area. As illustrated in Figure 4.4, the lens clips off all edges that pass the lens without actually connecting to a node within the lens. Such a local operation effectively clears the view, enabling the user to uncover and investigate edges within a local focus area. By dimming the visible context outside of the lens, the viewer's attention is subtly directed toward the lens focus.

The *layout lens* further supports the exploration of node neighborhoods, which are quite often not visible at a glance. To this end, the lens generates local neighborhood overviews by temporarily adjusting the graph layout based on the weighted distance between nodes and the center of the lens. Approaching a node with the lens and eventually centering the lens on top of it results in the complete neighborhood of that node being visible within the lens. When the lens is deactivated, all nodes return to their original position. Figure 4.5 illustrates the effect of the layout lens in contrast to the local-edge lens. While the local-edge lens allows the user to see edges better, the layout lens offers a better view on nodes.

Results and Discussion In summary, the presented intermediate-level interaction techniques solve common problems occurring when navigating and exploring graphs and they do so by utilizing properties unique to graphs.

Edge-based traveling makes use of the graph structure to ease navigation between connected nodes that are far apart in the layout. The radar view augments navigation by providing a useful outlook on what lies beyond the currently visible part of a graph's layout. Both techniques are *viewer-centric* in the sense that the users change their point of view onto the graph. The interactive lens techniques are *object-centric* as they shift the graph elements (i.e., nodes and edges) into the focus of the interaction. By taking advantage of the connectivity information of graphs, the lenses generate locally adapted views that grant access to information that is otherwise not visible at a glance.

With these interaction techniques specifically designed for graph data, we already see a connection to another key aspect of interaction, the tasks. For example, the edge-based traveling and the layout lens address tasks regarding path tracing and connectivity, which are relevant particularly in the context of graphs (Lee et al., 2006).

Tominski et al. (2009) show how the interaction techniques can be put to use in a larger graph visualization system to facilitate higher-level exploratory and analytic tasks. Informal feedback of users of this system indicates that the interactions are indeed helpful. There is also an online demo that gives the interested reader the opportunity to play with the graph lenses and the radar view (Tominski et al., 2015).

Interaction with graphs remains a topic of continued interest (von Landesberger et al., 2011; Wybrow et al., 2014). This is confirmed by recent advances, such as context-aware graph navigation (Ghani et al., 2011a), interactive graph matching (Hascoët and Dragicevic, 2012), novel off-screen techniques (Frisch and Dachselt, 2013), or navigation techniques for dynamic graphs (Bach et al., 2014).

For the next section, we maintain a data oriented view on interaction, but we make a switch from considering the data's *structure* to investigating the data's *frame of reference*, which will be spatio-temporal in our case.

Related Readings
Wong et al. (2003) • van Ham and van Wijk (2004) • Moscovich et al. (2009) • Ghani et al. (2011b) • Wybrow et al. (2014)

4.2.2   INTERACTING WITH SPATIO-TEMPORAL MOVEMENT TRAJECTORIES

A general goal of visualization is to show data in their frame of reference. Accepted methods exist for showing data in a spatial frame of reference (MacEachren, 1994; Kraak and Ormeling, 2010) and for visualizing data according to time (Aigner et al., 2011; Wills, 2012). Spatio-temporal movement trajectories are data for which both space and time are relevant. Existing approaches already address the difficulty of visualizing space, time, and movements simultaneously (Willems et al., 2009; Hurter et al., 2009; Krüger et al., 2013). Interaction plays an important role, because it allows users to explore different parts or different facets of the data.

In addition to space, time, and the movements themselves, there are also data attributes describing properties of movements (e.g., speed, acceleration, or sinuosity). An interesting question is how these data attributes can be visualized in the context of *where* and *when* the movements took place? A possible answer is to utilize the third dimension for the visual mapping (Ware et al., 2006; Grundy et al., 2009; Tominski et al., 2012b). The study by Amini et al. (2015) suggests that in such scenarios interactivity is even more important for comprehending 3D (and 2D) visualizations of movement data.

Let us take a closer look at Tominski et al.'s (2012b) hybrid 2D/3D visualization as depicted in Figure 4.6. The visualization design is based on stacking 3D bands along the third display dimension on top of a 2D map. This makes individual trajectories and their direction distinguishable, and data attributes can be color-coded along the bands. To facilitate maintaining

Figure 4.6: Movement trajectories visualized as 3D bands stacked on top of a 2D map.

the spatial context, the movements are additionally represented directly on a 2D map as color-coded 2D trace lines. 3D bands and 2D trace lines are linked through interactive highlighting. The visualization design further requires interaction support for two fundamental operations:

- **Spatial navigation:** Users must be enabled to explore different areas of the map and to look at the stack of trajectories (and the attribute values visualized there) from different perspectives.

- **Time integration:** As the visualization primarily shows the data in their spatial frame of reference, the link to the temporal dimension has to be established through interaction.

Approach   As we will see, dedicated context-dependent spatial navigation and lens-based dynamic queries are approaches that provide the required interaction support. Taken together, they account for the spatial and the temporal character of the data.

Context-dependent Spatial Navigation Jankowski and Hachet (2013) survey a number of 3D navigation techniques. While generic approaches, such as virtual trackballs (Henriksen et al., 2004), are generally useful, they do not consider the particular context of the scene being navigated. Context-dependent navigation is a combination of several basic navigation techniques. The question of which of them to apply and how to parametrize them is decided based on the current interaction context.

For spatial navigation, Tominski et al. (2012b) combine five different types of 3D navigation, including *viewer-centric* orbiting, fly-through, and look-around, as well as *object-centric* translating of the x-y map plane. The fifth technique is the *elevator*, which is a viewer-centric navigation constrained to the z-axis of the 3D scene. The elevator is similar to a scroll operation in that it takes the viewer to any level of the stack of trajectory bands.

A key to easing spatial navigation is to consider the context in which it takes place: Is a particular location on the 2D map of interest, is a specific point in a 3D trajectory band of concern, or can we assume a less focused interaction intent? A simple way to determine the context is to consult the position of the cursor in the visualization scene. Depending on the identified context, automatic adjustments can be made, including setting the center around which to orbit and correcting the speed with which to fly through the scene.

Even the already easy to use elevator can be further improved by considering the current context. When a user is investigating a 2D trace line on the 2D map, triggering the elevator is interpreted as the user's intent to visit the corresponding 3D trajectory band for closer inspection. Upon detection of this context, a shortcut takes the user directly to that trajectory band in the stack, no matter how high it may be and without any further manual operations.

With context-dependent navigation, users can easily explore the data's spatial frame of reference. How the temporal frame of reference can be integrated via an interactive lens will be explained next.

Integrating Time through an Interactive Lens   Trajectory bands and trace lines visualize *where* movements took place, but not *when*. Integrating time in full detail is hardly possible, because it conflicts with the existing visual design and would most certainly overload the display. A solution to this problem is to pick up the idea of interactive lenses, whose usefulness we already demonstrated in the previous section in the context of graph data. Now we use an interactive lens to provide on-demand access to temporal information.

The lens is basically an interactively adjustable spatial query region. Figure 4.7 illustrates its two modes of operation. The lens can be used to select movements that pass the 2D region defined by a lens circle (see Figure 4.7(a)). An additional dimension of control is provided by a lens cylinder (see Figure 4.7(b)), which enables the user to focus the selection on a certain range in the trajectory stack.

With a subset of movements selected with the lens, we can now visualize *when* the selected movements took place. To this end, an on-demand circular display is embedded into the visualization as illustrated in the right part of Figure 4.8. The outer ring shows temporally aggregated information in color-coded histogram bins. In the center is a scaled duplicate of the spatial query region, where individual trajectory points are displayed as dots. Links between the dots and the temporal scale at the inner ring establish a more direct connection between the space and time.

(a) 2D lens circle.

(b) 3D lens cylinder.

Figure 4.7: An interactive lens operating as 2D and 3D spatial query.

The particular visualization in Figure 4.8 reveals that movements in the selected region occur only on weekdays, but not on weekends as can be seen from the empty bins for Saturday and Sunday. The highlighted trajectory represents a movement that took place later on a Friday as indicated by the links (in darker gray) accumulating at the evening hours of the Friday bin.

By adjusting the query region and by switching between different temporal aggregations, the user can interactively control for which part of the data additional temporal information is displayed and how detailed the information will be.

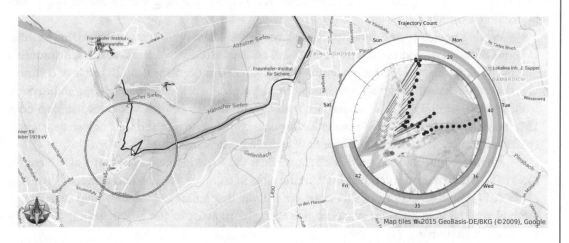

Figure 4.8: The lens (left) is linked to the display showing temporal information (right).

Results and Discussion   In summary, designing interaction techniques according to the frame of reference helps users in gaining a balanced understanding of the complex interplay of space, time, and data. Context-dependent spatial navigation facilitates exploring hybrid 2D/3D visualizations. Interactive lenses enable users to derive statements regarding the data's dependency on time in cases where the visualization is otherwise focused on the spatial dependency. Positive user feedback indicates that the interaction techniques are useful and usable: "The interaction is intuitive and works very much as expected." For an online demonstration, the interested reader is referred to Tominski et al. (2012c).

Yet for larger datasets, analytic methods are required to support the visual-interactive part. In their book, Andrienko et al. (2013) state that new methods of visualization need to be combined with new methods of algorithmic data analysis. The book picks up the solutions presented here and puts them to use in a larger visual analytics framework. Tominski et al. (2012b), Andrienko et al. (2013), and Andrienko et al. (2014) provide more details demonstrating the benefits of combining analytic, visual, and interactive means for exploring and analyzing movement data.

Related Readings
Krüger et al. (2013) • Roth (2013) • Jankowski and Hachet (2013) • Andrienko et al. (2013) • Amini et al. (2015)

SECTION SUMMARY

In this section, we elaborated on approaches taking into account the data as a key aspect of interaction in visualization. Looking at the data's *structure* and the data's *frame of reference*, we have seen how addressing the requirements imposed by the data (e.g., integrating space, time, and data attributes), on the one hand, and utilizing the characteristics of the data (e.g., navigating based on the structure of graphs), on the other hand, can help in designing and implementing useful interaction techniques.

While focusing on the data aspect, we have also touched other interaction-related aspects. In terms of tasks, we addressed path-tracing and connectivity-related tasks, which are unique to graphs. Behavior-related tasks are relevant when analyzing movement data. In terms of technology, the presented techniques address regular workspaces, where the multi-threading architecture from Section 4.1 and GPU acceleration are brought to bear to provide visual feedback quickly. In terms of the human, the *edge-based traveling* on a graph's structure and the context-dependent spatial navigation demonstrate the benefit of reducing interaction costs.

An interesting observation that can be made is the recurring pattern of lenses as interactive tools to support the user in carrying out specific tasks in complex visualization settings (Tominski et al., 2014). In later sections of this work, we will see that lenses are indeed a kind of universal method applicable in different scenarios, for example, to assist in the task of editing graphs as described in Section 4.3.2.

4.3   TASK-SPECIFIC INTERACTION TECHNIQUES

Interaction in visualization should not only address the data being visualized, but also the objectives that motivate the user to take actions. This brings us to the question how interaction in visualization can be related to the user's tasks.

Previous work on tasks in visualization is mainly concerned with designing the visual representations in a task-dependent way. However, not only the visualization design is dependent on tasks, but also the interaction design. In this context, Ware (2012) declares:

> "The optimal navigation method depends on the exact nature of the task."
>
> — Ware (2012)

While Ware's statement makes a point that interaction has to be task-specific at the intermediate level of navigation, it is clear that designing for the task at hand is beneficial at higher levels of interaction as well. In the following, we look at such higher-level interactions. Conceptually, they can be differentiated into interactions for:

- *consuming* data and

- *producing* data.

Consuming data means that data are the input to the human mind. Visual information seeking in general can be considered a high-level task during which data are consumed. For the first part of this section, we will focus on the specific task of visual comparison which also involves consuming data in various ways. We present a dedicated interaction method for visual comparison inspired by natural comparison behavior.

In contrast, producing data means that data are the output of the human mind. This is typically related to creating data models or editing data values. In the second part of this section, we study interaction for the task of editing graph structures. We describe a semi-automatic editing approach that combines interaction with automatic computation to facilitate inserting nodes into already complex graph layouts.

4.3.1   INTERACTION FOR COMPARISON TASKS

Visual comparison is a most relevant visualization task. It encompasses comparing multiple data values, groups of values, and also spatial regions or temporal intervals with specific data characteristics (Andrienko and Andrienko, 2006). Based on comparisons, users may formulate hypotheses about the data and draw corresponding conclusions, which indicates the high-level nature of the task.

Gleicher et al. (2011) underline the importance of comparison in visualization scenarios. Most of the solutions collected in their survey focus on supporting comparison by visual means, such as specific visual encodings or particular visual layouts. The survey also mentions interaction as an integral part, but often only rather basic interaction complements the customized visual

(a) Side-by-side comparison.   (b) Shine-through comparison.   (c) Folding comparison.

Figure 4.9: Natural behaviors of humans when comparing information on paper.

solution. Next we introduce a specifically designed and generally applicable interaction method for supporting visual comparison (Tominski et al., 2012a).

Approach   Visual comparison generally involves three phases. First, the pieces of information to be compared are identified and selected. Second, the selected pieces are arranged so that they are easy to compare. Third and finally, the actual comparison is carried out. These three phases must be supported by an appropriate interaction design. Particularly for the last phase it is interesting to study how humans naturally carry out comparison tasks. Figure 4.9 shows three fundamental behavior patterns that can be observed when humans compare information printed on paper:

- Side-by-side comparison: Sheets of paper are moved on a table until they are arranged side by side to facilitate comparison.

- Shine-through comparison: Sheets of paper are stacked and held against the light to let information shine through and blend.

- Folding comparison: Sheets of paper are stacked and individual sheets are folded back and forth to look at one or the other sheet in quick succession.

To support such comparison behaviors on the computer, the involved natural components and procedures have to be mimicked by virtual counterparts. In the first place, the workspace for the comparison has to be set up. We already mentioned the usefulness of the concept of zoomable user interfaces (see Section 2.3). Here it is utilized to create the virtual world in which comparison takes place. Natural sheets of paper are replicated as virtual visualization views that reside in the zoom space.

Now we can look at the virtual comparison procedures. The first phase is the selection of pieces to be compared. This is supported by enabling the user to mark interesting parts of a visu-

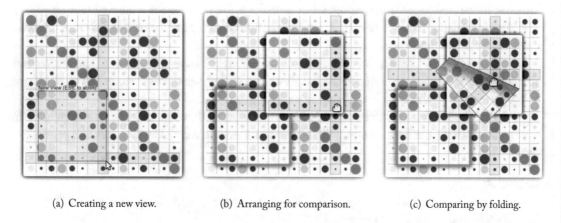

(a) Creating a new view.   (b) Arranging for comparison.   (c) Comparing by folding.

Figure 4.10: Phases of comparison exemplified with a matrix visualization.

alization view and to create new views of the marked parts. The second phase is the arrangement of the pieces to be compared. By moving views in the visualization space, the user can create any arrangement suitable for the later comparison. Snapping methods (Bier, 1988; Gleicher, 1995) can be employed to assist the user in arranging the views. The actual comparison is phase three. Through interactive arrangement, side-by-side comparison is already supported. Shine-through comparison can be realized via alpha-blending of stacked views. A specifically designed interaction also enables the user to perform the natural folding comparison in the virtual visualization space.

Figure 4.10 illustrates the three comparison phases for the example of a matrix visualization. For the actual comparison, the example uses the folding interaction. The figure also shows an additional visual cue that indicates where the smaller sub-matrix has been detached from its parent view.

Results and Discussion   The presented approach provides dedicated interaction support for the task of visual comparison. Each comparison phase is addressed by interaction techniques that draw inspiration from human behavior. Given their natural origin, it is almost certain that the interactions are intuitive and easy to use. This has been confirmed in a qualitative user study during which users expressed that the interaction "feels realistic" and is even "better than natural comparison."

An important characteristic of the approach is that it is generally applicable. All necessary operations can be implemented on the pixel stage of the visualization pipeline. This makes it possible to compare any kind of data representation, including visualizations of movements, dense pixel-based visualizations, or simply photos and screen captures.

At the same time, the approach is customizable to the particularities of the data. To this end, selected operations can be shifted to the data-oriented stages of the visualization pipeline.

This was illustrated here for matrix data, but is analogously possible for relational tabular data. It would be interesting to see data-specific customizations also for graphs or real 3D comparison of movement trajectories beyond the pixel level.

Tominski et al. (2012a) provides more details about the implementation of the interaction, about additional visual cues and interaction shortcuts, and about the user study. An online demo of the interaction techniques is available as well (Tominski, 2012).

With the interaction method for visual comparison, we have addressed a task for which the user *consumes* data. The next section will address a task during which the user acts as a *producer* of data.

Related Readings
Beaudouin-Lafon (2001) • Dragicevic (2004) • Isenberg and Carpendale (2007) • Valli (2008) • Gleicher et al. (2011)

4.3.2   INTERACTION SUPPORT FOR EDITING TASKS

Visualization users U typically engage in tasks where they consume data D through exploring visual representations V with the goal to gain insight (van Wijk, 2005), or short $U \leftarrow V \leftarrow D$. Yet, there are situations in which users have to produce or manipulate data, for example, to insert missing items, update erroneous data values, or delete obvious outliers. Such data editing tasks can be supported through means of visualization as well (Baudel, 2006). The goal is to use visual representations as a bi-directional mediator between data and users, or short $U \leftrightarrow V \leftrightarrow D$.

From an interaction point of view, there is a gap between visual data exploration and editing. Visual exploration focuses on interactive control of the visualization process $U \rightarrow V$, as documented several times in the previous sections. Data editing $U \rightarrow D$ is usually a manual, cumbersome, and time-consuming process (Kandel et al., 2011). In general, $U \rightarrow V$ and $U \rightarrow D$ are separate tasks that are carried out with separate tools. This not only disrupts the user's workflow, but also requires costly mental context switches, which make the whole procedure prone to human error.

Narrowing the gap between exploration and editing, we next present an approach that integrates both aspects with an emphasis on the editing part (Gladisch et al., 2014). As an example for the data to be edited we focus on graphs with customized layouts, more specifically on the graph structure. Because editing such graph layouts can be quite complex, the otherwise purely interactive editing is supported by automatic methods.

Approach   The visual basis of the approach is an orthogonal node-link representation. It shows the structure of the graph, textual labels for graph nodes, and optionally data attributes via color-coding. The node-link representation is again embedded in a zoomable space that provides the necessary interaction for exploring the graph.

To enable the user to edit the graph's structure and its layout, an interactive lens, called *EditLens*, is provided. In the previous sections, we considered interactive lenses mainly as tools to

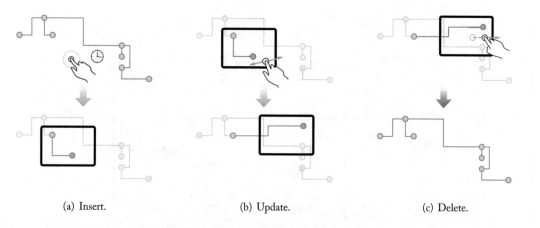

(a) Insert.   (b) Update.   (c) Delete.

Figure 4.11: Illustration of the EditLens being applied to insert, update, and delete a node.

provide alternative visual representations of the data (see the lenses for graphs and movement data from Section 4.2). In contrast, the EditLens is a tool to edit the data, that is, to insert, update, or delete data items. For graphs, the data items to be edited are nodes and edges. Figure 4.11 illustrates editing operations for a node.

Inserting a new node or edge into an already existing complex graph layout is usually a demanding task. In our particular scenario, the graph layout is custom-made and obeys certain application dependent constraints (e.g., specific types of nodes must be located in dedicated regions of the layout). Editing such graphs requires finding adequate positions for nodes to be inserted and may also involve routing edges through the already established complex layout.

The key idea of the EditLens is to ease editing tasks by relieving the user of defining precise points for nodes or edge routes. Instead, the user specifies just the local region where an edit operation is to take effect. In other words, *point-wise* manual editing is relaxed to *region-wise* semi-automatic editing via the EditLens. While the interactively adjustable lens region acts as a coarse solution specified by the user, precise positions and edge routes are computed by automatic methods as described next.

The first step is to determine a suitable unoccupied area within the EditLens where an edited item can be placed. The precise spot within that area is computed based on different heuristics (Gladisch et al., 2014). These heuristics prioritize different graph aesthetics criteria, for instance, maximum distance to other nodes, short overall edge length, or low number of edge bends. During the editing, the user can freely choose which heuristic to apply depending on the situation at hand. The EditLens will suggest suitable positions of nodes and edge routes accordingly. When the user agrees with a suggested solution, the result of the edit operation is committed to the data. In cases where the EditLens can find only insufficient or no solutions due to conflicting constraints, manual refinement is still possible with classic point-wise editing.

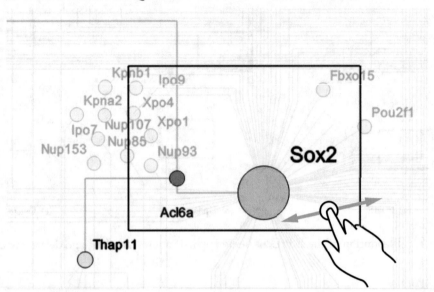

Figure 4.12: Using the EditLens for inserting node *Acl6a* into a complex biological network.

The semi-automatic editing has to be supported by appropriate visual feedback. The fact that the EditLens is in operation is indicated by a black rectangle. Only the nodes and edges being affected by the ongoing edit operation are shown fully saturated, whereas all other nodes and edges are dimmed. If the EditLens cannot provide valid suggestions for the edit operation, the lens frame turns red to notify the user.

Results and Discussion   The EditLens has been applied to a real-world problem where bio-informatics scientists maintain a network of genes and biological relations among them. As new genes or relations are discovered and reported in the literature, the network is edited to include the newly available information. Figure 4.12 shows a small part of the network. One can easily imagine how cumbersome and time-consuming manual editing of the network can be.

In a small qualitative study with two experts and two non-expert users, the manual editing has been compared with the semi-automatic editing using the EditLens (Gladisch et al., 2014). The collected user feedback suggests that "the EditLens is very useful and can clearly reduce the editing effort" and that "the automatic suggestion of node positions and edge routes is obviously beneficial." We can conclude that the EditLens significantly narrows the gap between data editing and data exploration in the sense of $D \leftrightarrow V \leftrightarrow U$.

Editing node-link diagrams is a research challenge in its own right (Frisch, 2012). The concrete implementation of the EditLens focuses on the structure of graphs with orthogonal layouts. This leaves two key questions. First, how can editing be supported not only in terms of the graph structure, but also with respect to data values associated with nodes and edges? Second, are there other visual representations than node-link diagrams that are suitable alternatives to improve specific editing task? For example, are matrix representations better suited for editing the edges of a graph? Finding answers to these questions will help to further narrow the exploration-editing gap.

Related Readings
Baudel (2006) • Dwyer et al. (2008) • Frisch et al. (2009) • McGuffin and Jurisica (2009) • Kandel et al. (2011)

SECTION SUMMARY

In this section, we have shown how visualization tools can be designed with respect to the task aspect of interaction. We considered high-level interaction to support two fundamentally different types of tasks: tasks that involve the consumption of data and tasks that are related to the production of data. As concrete task instances we studied visual comparison and data editing. We have seen that a key to success is to pay attention to the procedures and workflows behind tasks.

We can again see connections to the other key aspects of interaction. The EditLens has been designed specifically for graphs, which confirms the importance of the data aspect. While the approach for visual comparison is generally applicable in terms of the data, we also mentioned that specific customization to particular data characteristics can be useful. An obvious link to human aspects is the interaction design based on natural comparison behavior. By following natural comparison strategies, a solution could be obtained that is intuitive and effective. Also the EditLens adheres to a user-centered design insofar as it supports the editing workflow of real users working on real problems.

Speaking in terms of technology, we began to make a shift from classic mouse and keyboard interaction for the comparison approach to multi-touch interaction for the EditLens. Touch interaction has the advantage that users can directly manipulate the data under their fingertips. Typical precision problems with touch interaction do not surface for the EditLens, because it is a region-oriented approach and precision is taken care of by automatic methods. This already hints at the importance of considering technological aspects when designing interaction solutions. In the next section, we further explore the utilization of modern interaction technology in the context of visualization.

4.4 UTILIZING MODERN TECHNOLOGY FOR INTERACTION

With data-specific and task-centric interaction as dealt with in the two previous sections, we have covered the core of interaction in visualization. As illustrated in Figure 3.1 back on page 25, this core is flanked by the technology on the one side and the human on the other side. Both of these flanks will be addressed in the following. We consider the technology in this section and the human in the next Section 4.5.

Undeniably, the technology plays a key role in interactive visualization, because it provides the facilities to display visual representations and the means to interact with them. Typically, the technical setup for visualization applications is a regular desktop computer to which a regular display as well as a mouse and a keyboard are connected. In this section, we go beyond mouse and keyboard interaction and regular displays.

Addressing the technology aspect typically involves studying interaction at the lower level of basic ways of transmitting interaction intents from the human user to the computer. There are different interaction modalities that all come with their own individual strengths and weaknesses. It is beyond the scope of this section to comprehensively discuss all possible options. Our considerations are focused on two principal ways of establishing an interaction channel between the human user and the computer:

- *tracked objects* – users manipulate objects that are tracked and

- *tracked humans* – users themselves are tracked.

While mouse and keyboard are the classic input devices, our goal is to illustrate alternative means of using tracked objects for interaction. To this end, we discuss *tangible views* as a novel way of interacting with visual representations of data. At the same time, tangible views are a novel form of lightweight displays that offer new possibilities for visualization applications. As an example where users are tracked, rather than objects, we look at *proxemic interaction* in front of large displays. In particular, we investigate how tracking the user's position and viewing direction can support graph exploration on a large tiled display wall.

4.4.1   TANGIBLE VIEWS FOR INTERACTION AND VISUALIZATION

Direct manipulation as advocated by Shneiderman (1983) is a most prevalent theme of interaction in visualization. Many visualization techniques are designed so as to allow the user to directly manipulate the visual representation or the underlying data.

Yet, often the term *direct* just means that manipulations affect the visual object directly under the pointer, which is a rather limited interpretation of directness. In fact, standard interaction is quite indirect: the pointer is typically controlled with a mouse, whereas visual feedback is shown on the display. Researchers have started to explore how truly direct touch interaction can effectively support visualization (Voida et al., 2009; Kosara, 2011; Kim and Elmqvist, 2012). Next, we expand basic touch interaction *on* a display by tangible interaction *with* the display as suggested by Spindler et al. (2010).

Approach   The starting point of the approach is a horizontal touch-enabled tabletop device. The tabletop serves to visualize the data and to receive touch input from the user. In this setup, visualization and interaction remain in the horizontal two-dimensional tabletop plane. The idea now is to utilize the three-dimensional space above the tabletop to provide enhanced visualization and interaction functionality.

To this end, the concept of the *PaperLens* (Spindler et al., 2009) is extended to what is called *tangible views*. Tangible views are lightweight "devices" that act as additional displays in the space above the tabletop (see Figure 4.13). In a most simple instantiation, a tangible view can be a piece of cardboard onto which a projector transmits visualization content, in which case the display is passive. A tangible view can also be active, that is, it is capable of displaying graphical content on its own without the help of an external projector, for example, a tablet device.

Irrespective of being passive or active, the key characteristic of tangible views is that they are spatially aware. Through constant tracking of the tangible views, the system always knows their position and orientation in three-dimensional space. This opens up new possibilities for interaction. The extended interaction vocabulary of tangible views includes basic translation and rotation in three dimensions as well as gestures of flipping, tilting, and shaking. By providing tangible views that are distinguishable by shape or appearance it is possible to create an interaction toolbox, where users can infer interaction functionality from the look of a tangible view. Multiple tangible views can be applied simultaneously for advanced interaction and for adding display space for visualization purposes.

Providing an extended interaction vocabulary is only one part of the approach. The second part is to appropriately map the vocabulary to typical interactions in visualization scenarios. Here, we take a brief look at an application of tangible views for visual comparison of graph data. We

Figure 4.13: Tangible views in use for comparing data matrices.

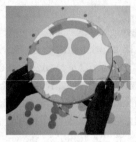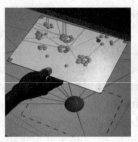

Figure 4.14: Tangible views applied to different types of visualizations.

chose this example as it is interesting to see the novel way of tangible interaction for visual comparison in contrast to the visual comparison approach based on classic mouse interaction from Section 4.3.1.

How visual comparison can be carried out using two tangible views is depicted in Figure 4.13. The tabletop shows a color-coded graph matrix as the background visualization. In order to select two sub-matrices to be compared, the user moves two tangible views horizontally above the tabletop. A freeze gesture is carried out to fix the selection. This allows the user to arrange both tangible views side by side for closer inspection and comparison. Once the views are sufficiently close to each other, the system recognizes the user's comparison intent and indicates by a colored frame (green in our case) the overall aggregated similarity of the sub-matrices. Shaking the tangible views releases the freeze and the user can start over with selecting other parts of the matrix.

As illustrated by the visual comparison application, tangible views offer a novel and intuitive way of interacting with visual representations of data. This is true not only for comparison tasks and graph data, but for a broad range of visualization problems. Spindler et al. (2010) apply tangible views in a number of case studies, including sampling in parallel coordinates, fisheye magnification in scatter plots, multi-level exploration of hierarchical graphs, and spatio-temporal data analysis with space-time cube (see Figure 4.14). These case studies further explore and demonstrate the usefulness of the introduced extended interaction vocabulary.

Results and Discussion   From a conceptual point of view, tangible views led to three key results related to interaction and visualization. First, tangible views integrate display and interaction device. The user interacts directly *with* the tangible view to satisfy an interaction intent. Interaction *on* tangible views or the tabletop display can provide additional functionality. Second, tangible views enhance common 2D interaction with tangible 3D interaction above a tabletop display, thus extending the typical interaction vocabulary for visualization. Third, the enhanced interaction vocabulary and extended physical display space allow us to create a tangible experience of otherwise purely virtual visualization techniques and concepts, including overview+detail, focus+context, and coordinated multiple views.

With great power comes great responsibility. While the mentioned case studies illustrate how the extended expressive power of tangible views can be utilized, it remains to be studied what will be good interaction designs in other situations. The case studies also indicate that good designs require taking into account the concrete data and the specific task of the user.

The approach of tracking tangible views in 3D also raises questions regarding precision of the interaction and regarding fatigue of users operating multiple tangible views. Investigating these issues requires extensive user studies. First results in this direction have been published by Spindler et al. (2012, 2014). They indicate that tangible spatial interaction is indeed a promising alternative when working in and with layered zoomable information spaces, which are common in visualization scenarios.

By the example of tangible views, we illustrated the usefulness of novel interaction technology where users manipulate tracked objects. Next we describe how tracking the user, rather than objects, can support visualization on large high-resolution displays.

Related Readings
Ishii and Ullmer (1997) • Ullmer et al. (2003) • Holman et al. (2005) • Kim and Elmqvist (2012) • Jackson et al. (2013)

4.4.2   PROXEMIC INTERACTION FOR WALL-SIZED VISUALIZATION

The approach from the previous section visualized data on tangible views and on a tabletop device. These and other output devices with conventional pixel resolution are typically limited in the amount of information that can be displayed. Thanks to technological advances, large high-resolution displays are now becoming available to a broader range of users. The larger physical size and the increased pixel resolution offered by such displays have obvious advantages for visualization applications. Particularly in light of big data, being able to visualize much more information is an exciting prospect.

Yet with larger size and more pixels there also come new challenges. New solutions are needed to support the visualization and interaction on wall-sized displays. Andrews et al. (2011) point out that

> "Replacing the conventional monitor with a large, high-resolution display creates a fundamentally different environment that is no longer defined purely in terms of the technical limitations of the display, creating a new collection of design opportunities, issues, and challenges."
>
> — Andrews et al. (2011)

Here, we look at a scenario where data are visualized on a tiled display wall and focus on the question of how to interact in such a setting. *Proxemic interaction* has been identified as a promising answer to this question (Ballendat et al., 2010). We describe such an interaction design where physical navigation in front of a display wall steers the exploration of hierarchical graphs at different levels of detail (Lehmann et al., 2011).

Figure 4.15: Exploring a graph on a large high-resolution display wall.

Approach As shown in Figure 4.15, the tiled display wall consists of 24 individual displays covering an area of 3.7 m × 2.0 m with a total resolution of 11,520 × 4,800 which amounts to 55 million pixels. The graph being explored is a hierarchical graph with different levels of abstraction. Such graphs can be visualized using standard node-link representations. Nodes can be expanded to get a finer view on the data or collapsed to get a coarser representation. In standard desktop applications, individual levels of abstraction are typically accessed via mouse and keyboard interaction (e.g., clicking on nodes to expand them and holding down a key while clicking to collapse them). In our scenario, this is obviously impractical due to the large distances that would need to be covered with the mouse cursor across multiple displays.

An alternative is to control the exploration of the graph by the user's physical movement in front of the large display wall. For our particular example, head tracking is used to acquire information about the user's position and orientation (6 degrees of freedom). The tracking information is utilized to implement two interaction techniques: the *zone technique*, which can be combined with the *lens technique*.

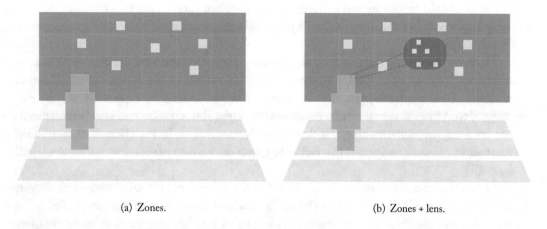

(a) Zones.   (b) Zones + lens.

Figure 4.16: Schematic illustration of the zone technique and the lens technique.

For the zone technique, the space in front of the display wall is subdivided into multiple zones with increasing distance to the display (see Figure 4.16(a)). Each zone corresponds to a level of abstraction of the graph. When the user moves toward the display, the graph is visualized at greater detail. This approach corresponds to natural human behavior. When interested in details, humans typically step up to the object of interest to study it in detail. In order to obtain an overview, the user can step back. Stepping backward into zones farther away from the display automatically adjusts the visualization to show higher levels of abstraction.

While the zone technique can be used to globally control the level of abstraction, the lens technique has been designed to enable the user to access details for local parts of the graph. To this end, the tracking information about the orientation of the user's head is utilized to estimate where the user is looking. Based on this estimation, an interactive lens is positioned on top of the regular graph visualization (see Figure 4.16(b)). Nodes that are inside the lens are automatically expanded to reveal more detailed information. The lens technique enables the user to scan the graph in a focus+context fashion by simply moving the head around. Filtering the tracking input and smoothly animating node expand and collapse operations help to avoid flickering caused by natural head tremor and to maintain a reasonably stable visualization.

The user's distance to the display can be used not only to control the level of abstraction, but also to derive a suitable labeling of the graph. A user standing close to the display is presented with more and smaller labels for individual nodes. When looking from a greater distance, the user will see fewer, but larger labels for clusters of nodes.

Results and Discussion   The interaction techniques have been tested in a pilot study (Lehmann et al., 2011). The zone technique was reported as the approach that is easier to use, but on the other hand, the lens technique offered more control over where increased detail is to be shown. Overall, the proxemic interaction through physical movement in front of a display wall proved to be a valuable alternative in cases where classic means of interaction fail. Physical movement not only better matches the scale of the display, it also corresponds with natural interaction with real-world objects.

Here, we considered the relatively simple task of adjusting the level of abstraction of a graph visualization. Jakobsen et al. (2013) study proxemic interaction in the context of visualization in more detail. Their results suggest that zooming tasks can indeed be supported well by physical navigation. Other research results indicate that physical navigation can also be beneficial for higher-level analytic sensemaking (Andrews and North, 2013). In the context of multi-display environments, physical navigation can facilitate group discussions of visualization results (Radloff et al., 2015).

In terms of the visualization itself, we relied on rather basic standard node-link diagrams. There are other visualization designs that offer a better adaptation to the properties of large displays. A particularly interesting example is the hybrid-image visualization by Isenberg et al. (2013a). A hybrid image supports multi-scale viewing for multiple users collaborating in front of a large display.

Related Readings
Yost and North (2006) • Ball et al. (2007) • Keefe et al. (2012) • Greenberg et al. (2014) • Jakobsen and Hornbæk (2015)

SECTION SUMMARY

In this section, we presented two technology-oriented approaches to interaction in visualization. The first was based on tracking tangible views above a tabletop display, the second on tracking the user in front of a large display wall. Tangible views significantly extend the interaction vocabulary for visualization applications. We are no longer limited to interaction *on* the display, but can now interact *with* the display. Tangible views also offer more space for displaying information. Even more space is available on mega-pixel display walls. In such scenarios, we have to match the scale of the interaction with the scale of the display. Physical navigation as a form of proxemic interaction turned out to be well-suited for exploring graphs at different levels of abstraction. Tangible views and proxemic interaction illustrate the potential of going beyond regular displays and standard interaction. This topic is discussed further in studies by Keefe and Isenberg (2013), Jakobsen et al. (2013), Isenberg et al. (2013b), and Jansen and Dragicevic (2013).

While focusing our attention on the aspect of technology, we could once more see links the to other key aspects of interaction in visualization. That the human is central in both approaches is self-evident. The data and the tasks were considered in particular for the different case studies of tangible views. Each study focused on specific data and tasks. One exemplary use case for tangible views was multi-level exploration of hierarchical graphs. The very same data and task were addressed when we were studying physical navigation in front of a display wall. The techniques described earlier in Section 4.2.1 support multi-level exploration of hierarchical graphs as well, but they rely on classic mouse interaction.

A question that remains largely unexplored so far is which technology, tangible views, physical navigation, or classic mouse interaction, fares best for different types of data and different tasks. A good starting point for investigations in this direction would be a comparative study with the example of multi-level graph exploration. But already setting up the environment for such a controlled study would pose considerable challenges, technology-wise and evaluation-wise. Yet the results of such a study would help us develop an understanding of which technologies are best suited to support the human user. In the next section, we will address the human user specifically by integrating automatic and interactive methods.

4.5   SUPPORTING HUMAN INTERACTION WITH AUTOMATIC METHODS

We now come to the final aspect of our interaction-oriented view on visualization, the human. The human user is the source of interaction intents transmitted to the system via different interaction modalities. At the same time, the human user is the sink for visual information conveyed through display technology.

Developing interaction with a focus on the human user involves addressing aspects of fluidity, naturalness, and cost efficiency of the interaction. Actually, all these aspects have already been covered in the previous Sections 4.2–4.4. The easy navigation based on graph structures, the context-dependent 3D navigation of trajectory data, the interaction for visual comparison inspired by natural behavior, the data editing approach that follows real-world users' workflows, the approach that makes interaction for visualization tangible, and the natural physical navigation in front of a large display have all been designed with the human user in mind.

In this section, we continue to address the human user, this time by reducing the costs involved when users interact. Lam (2008) attributes interaction costs in visualization to the two gulfs in Norman's (2013) action cycle:

- gulf of execution—carrying out the interaction and

- gulf of evaluation—understanding the visual feedback.

The following paragraphs present two approaches that aim to narrow these gulfs. Yet in contrast to the previous sections where we focused on interactive solutions, we will now consider *automatic methods* to assist in interactive visualization. In terms of the gulf of execution, we look at event-based concepts to automate certain actions that otherwise would need to be performed manually. In terms of the gulf of evaluation, we address typical questions arising when users navigate larger information spaces. We describe a degree-of-interest (DOI) approach that automatically generates navigation recommendations to help users making informed decisions on where to navigate next.

4.5.1   REDUCING INTERACTION WITH EVENT-BASED CONCEPTS

Interaction in visualization has been advocated throughout this work as a means to empower the human user to control the visualization as needed for the data and the tasks at hand. Yet in Section 2.5, we also indicated that interaction is not a universal cure and that interaction can be a burden to the human user. So it makes sense to critically question all interactive input that visualization tools solicit from the user. After all, it is the task of the visualization to present relevant information effectively and expressively, and not primarily the task of the human user to parameterize the visualization appropriately to achieve this.

Thinking about reducing user interaction to a reasonable and useful minimum implies that we also have to look for alternative sources of input to be able to derive visual representations that still reflect the user's needs. One such source of information is the data itself and patterns of interest residing in the data. Knowing that certain parts of the data are of special interest, one can automatically trigger adjustments of the visual representation to emphasize these interesting parts without the need of user interaction.

Approach   In order to automate the adjustment of visual representations according to user interests, visualization can be complemented with event-based methods (Tominski, 2011). As illustrated in Figure 4.17, such an event-enhanced visualization comprises three stages: event specification, event detection, and event representation.

The event specification is concerned with defining event types that formalize the notion of "interesting parts of the data." Predicate logic formulas are utilized to express three different kinds of event types addressing the relational data model. The user can specify tuple event types (e.g., tuple values exceed a certain threshold) and attribute event types (e.g., data attribute with the highest value). In order to deal with changes over time, there is also support for sequence event types (e.g., sequence of days with rising temperature). Composite event types can be compiled using set operators.

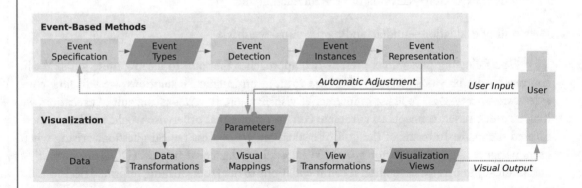

Figure 4.17: Model of event-based visualization.

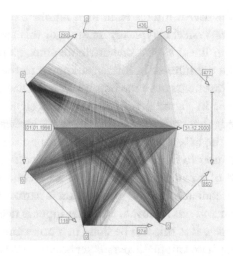

(a) Default parameterization.

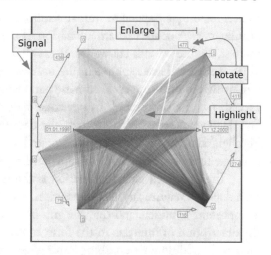

(b) Improved parameterization.

Figure 4.18: Automatic event-based adjustment of the TimeWheel.

The second stage is the event detection. At this stage, the data are searched for matches of the interests expressed via event types. Different algorithms are involved at this stage, including query mechanisms for relational databases as well as search algorithms for sequence patterns. Efficiency of these algorithms is of importance, because the search has to be carried out upon every change of the data. This is particularly critical for dynamic data streams that undergo frequent changes. Once the event detection reports matches in the data, actual event instances are created. Event instances bear three important pieces of information: (1) the fact that something interesting has been found in the data, (2) where in the data interesting parts are located, and (3) what made these parts interesting (i.e., the event type). This information is the input to the event representation.

The goal for the event representation is to automatically adjust the visual representation so as to (1) communicate the fact that something interesting is in the data, (2) emphasize interesting parts among the rest of the data, and (3) indicate the event type, where possible. These goals are to be achieved by re-parameterizing the visualization. Therefore, it is a necessary requirement that the visualization provides an appropriate set of visualization parameters. If this requirement is satisfied, the event-based approach automatically triggers the execution of instantaneous or gradual parameter changes upon the detection of events. These automatic parameter adjustments reduce the need for manual interaction.

Results and Discussion   Tominski (2011) describes exemplary event-based enhancements for several visualization techniques, including the Table Lens (Rao and Card, 1994), the space-time cube (Kraak, 2003), and the TimeWheel (Tominski et al., 2004). Figure 4.18 illustrates the Time-

Wheel technique applied to human health data. The user is interested in high numbers of cases of influenza. A default parameterization of the TimeWheel is typically unaware of this specific interest. Figure 4.18(a) shows influenza (light green lines) as just one attribute among others. With the event-based approach, the user can make the interest in influenza known to the system, which in turn triggers automatic parameter adjustments once the number of influenza cases exceeds a certain threshold. In our concrete example in Figure 4.18(b), a red frame signals the fact that events have been detected. The data records that correspond with the user's interest are highlighted and the TimeWheel is rotated so that the axis representing influenza is on the top. Enlargement of the influenza axis further helps the user to focus on the interesting cases.

The improved parameterization is the result of an automatic reaction to an event and as such does not require any user interaction. As a consequence, the gulf of execution is narrowed. It is important to understand that the gulf is narrowed and not removed. Still user input is needed: the description of interests in the form of event types. But this can be done in a one-time pre-process. In addition to describing event types, appropriate automatic parameter adjustments must be defined, which is typically a task for the visualization designer. An interesting question would be to derive rules or guidelines on what visualization parameters are needed and how they can be adjusted in order to obtain appropriate visual effects for the event representation.

We have seen that event-based concepts can be a useful complement to interaction in visualization. Automatic reaction to user-defined events of interest can reduce the gulf of execution. An approach to narrow the gulf of evaluation will be presented in the next section.

Related Readings
Horvitz (1999) • Allen et al. (1999) • Healey et al. (2008) • Keim et al. (2008) • Bouali et al. (2015)

4.5.2   NAVIGATION RECOMMENDATIONS FOR INFORMED INTERACTION

The gulf of evaluation relates to the costs arising when interpreting the visual feedback resulting from interacting with the visualization. Particularly during data exploration, evaluation costs might accumulate, because exploring the data generally means carrying out a number of interactive navigation steps, which all ensue evaluation costs (Furnas, 1997). Consider, for example, the interaction techniques for navigating along the structure of graphs as presented in Section 4.2.1. After taking a navigation step, the user has to evaluate the updated visual representation of the graph structure. Typical questions that users might ask themselves include: "Where am I in the structure?", "What structural patterns can I see here?", and "How is what I see related to what I have seen before?"

Spence (2007) further lists the questions: "Where can I go?", "What lies beyond?", and "Where can I usefully go?" The goal now is to support the user in answering these questions by automatically computing and presenting *recommendations* to the user. Recommendations have already proved useful for supporting users in selecting visualization approaches (Gotz and Wen,

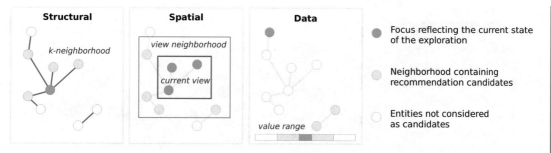

Figure 4.19: Neighborhoods for collecting recommendation candidates.

2009). Here we consider recommendations to reduce evaluation costs during interactive navigation in hierarchical graphs (Gladisch et al., 2013).

There are two types of navigation for hierarchical graphs. *Horizontal navigation* relates to navigation in the graph layout. Typical operations to adjust the view on the graph layout are scrolling and panning also in combination with zooming. *Vertical navigation* denotes navigation in the data, more precisely, navigation between different levels of abstraction of the data. The level of abstraction can be set globally, but also local adjustments are possible by expanding or collapsing individual clusters of nodes (Elmqvist and Fekete, 2010). The question is what are useful horizontal or vertical navigation steps to be recommended to the user?

Approach   The general procedure is as follows. First, a set of graph nodes is determined to define recommendation candidates. Second, the candidates are ranked and the top-ranked nodes are selected as actual recommendations. Finally, the selected recommendations are communicated visually to the user.

For building the set of recommendation candidates, Gladisch et al. (2013) consider the neighborhood of the current exploration situation, that is, the context of the visualization content currently visible on the display. As illustrated in Figure 4.19, the neighborhood can be defined in three different ways: structural neighborhood, spatial neighborhood, and data neighborhood. The structural neighborhood is based on the k-neighborhood of the graph. The spatial neighborhood is defined in terms of node positions in the graph layout. A data neighborhood can be determined based on similarity among the attribute values associated with the graph nodes. Restricting the recommendation candidates to a local neighborhood around the user's focus has two advantages. First, the candidates are guaranteed to be related to the part of the graph currently being visible. Second, the neighborhood is much smaller than the dataset as a whole, which eases the ranking of the candidates.

To determine where the user can *usefully* go, a definition is needed of what *useful* means. An effective concept in this regard is the degree of interest (DOI). The degree of interest is typically defined through a DOI function. The literature describes different components of DOI functions,

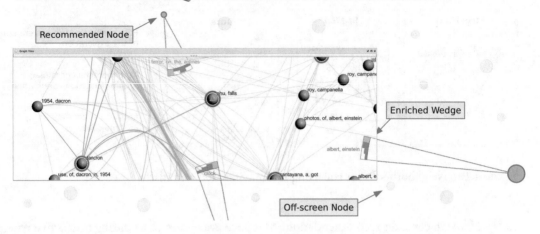

Figure 4.20: Navigation recommendations indicated by rings and enriched wedges.

including API (a priori interest), UI (user interest), DIST (distance to focus), and KNOW (interest degradation for already visited data). All components can be weighted to fine-tune the DOI computation depending on the application context. With an appropriate DOI specification, one can automatically compute and assign DOI values to all recommendation candidates and rank them accordingly. The top-ranked candidates are selected as actual navigation recommendations to be presented to the user.

Navigation recommendations are additional pieces of information that need to be conveyed to the user in addition to communicating the actual data. Corresponding visual cues have to be designed so that they subtly indicate the recommended target nodes. For horizontal navigation, target nodes can be on-screen or off-screen. As shown in Figure 4.20, on-screen targets can be marked with rings, whereas off-screen destinations are hinted at by visual cues called *enriched wedges*. On demand, the visual cues can visualize the DOI values assigned to the recommended nodes. For vertical navigation, targets are by definition located at levels of abstraction different from the current level, and hence they are definitely not visible. Therefore, recommendations for vertical navigation are attached to anchor nodes whose expansion (or collapse) would bring the recommended target to the display. Subtly pulsing rings around anchors suggest that an expand (outward pulsing) or a collapse (inward pulsing) operation will uncover a target of interest.

Results and Discussion   Automatically computed navigation recommendations assist the user in making informed navigation decisions. Thanks to the DOI concept, we are able to recommend targets that reflect the user's interest, provided that the DOI components are appropriately defined and weighted. The design of the visual cues that hint at the recommended targets has to follow a defensive strategy in order to only minimally interfere with regular data exploration. Only when users have difficulties in determining a good next navigation target on their own should their attention shift to the recommendations.

A proof-of-concept implementation has been tested with different hierarchical graphs of moderate size (hundreds of nodes and thousands of edges). The tests indicate that the approach in general is technically feasible (Gladisch et al., 2013). Navigation recommendations could be extracted from the graphs on-the-fly, while the computations were not hindering any regular exploration activities of the user.

Still a difficulty is to appropriately outfit the DOI function. The components API, UI, and DIST can be defined following van Ham and Perer (2009). However, the definition of the KNOW component to capture what parts of a graph have already been explored is rather rudimentary. A promising alternative could be to use gaze-controlled approaches, for example, based on the work by Okoe et al. (2014).

Related Readings
Furnas (1986) • Gustafson et al. (2008) • Crnovrsanin et al. (2011) • May et al. (2012) • Abello et al. (2014)

SECTION SUMMARY

In this section, we investigated how the human user can be supported when interacting with visual representations of data. With the event-based visualization and the navigation recommendations, we presented two approaches to reduce the gulfs of execution and evaluation. Interestingly, both approaches are, in a sense, in contrast to our previous focus on interactive solutions, because at their core they are automatic methods: finding interesting events in the data and determining interesting navigation targets. What this contrast illustrates quite well is that it definitely makes sense to critically question interaction and investigate alternative methods to improve visualization.

Of course complementing human interaction with automatic computation raises the question of proper balancing of automatic and interactive methods. This question has to be answered individually depending on the application scenario and user expertise. Cooper et al. (2007) state that casual users need basic functionality, that experienced users tend to explore enhanced functionality, and that expert users seek ways to automate tasks. Although these statements provide some general indication, it is left for future work and longitudinal studies to more thoroughly investigate guidelines on balancing interaction and automated methods in visualization settings.

A second concern, not only relevant here, is the need to appropriately estimate user interest. Both presented approaches depend on the existence of suitable definitions of user interest: event types and DOI specification. Yet in real-world applications, these definitions might turn out to be difficult to set up. Therefore, it makes sense to continue researching alternative means for estimating user interest, where gaze-based approaches appear particularly promising.

Overall, we can conclude that addressing the human is vitally important for interaction in visualization. For the future, we see an increased relevance of complementing interactive visualization with assistive methods. Under the umbrella of guidance in visualization, ongoing research investigates how the human can be supported at different levels, including guidance to visualize data effectively, guidance to assign tasks to the right user, and guidance to employ suitable technologies (Schulz et al., 2013b).

4.6   SUMMARIZING REMARKS

In this chapter, we presented 1+8 approaches related to the key aspects of interaction in visualization as identified in Chapter 3. All methods and techniques were described in a compact fashion focusing on important questions and corresponding answers. As summarized in Table 4.1, we started out with a multi-threading architecture laying out a technical foundation for implement-

Table 4.1: Summary of approaches in relation to data, tasks, technology, and human

	Data	Tasks	Technology	Human
Section 4.1 **Architecture**	Generic	Generic	Multi-core processors	Efficient interaction loop, circumvent multi-threading pitfalls
Section 4.2.1 **Data** Data structure	Graph data	Navigate, explore	Desktop	Enhanced navigation and exploration
Section 4.2.2 **Data** Data frame of reference	Movement data	Navigate, explore	Desktop	Enhanced navigation and exploration
Section 4.3.1 **Tasks** Consume data	Generic	Compare	Desktop	Interaction inspired by natural human behavior
Section 4.3.2 **Tasks** Produce data	Graph data	Edit	Desktop, touch interaction	Interaction based on real-world work-flow and supported by automatic means
Section 4.4.1 **Technology** Tracked objects	Generic	Generic	Tangible interaction, touch interaction	Truly direct interaction with and on display
Section 4.4.2 **Technology** Tracked human	Graph data	Explore	Proxemic interaction, wall-sized displays	Physical navigation, natural interaction
Section 4.5.1 **Human** Gulf of execution	Generic	Encode, reconfigure	Generic	Automatic visualization parametrization
Section 4.5.2 **Human** Gulf of evaluation	Graph data	Navigate	Generic	Recommendation of navigation

ing interactive visualization. Then, the data aspect, the task aspect, the technology aspect, and the human aspect were individually addressed and illustrated by two approaches each. We presented techniques for intermediate-level interaction with graph data and movement data. Higher-level interaction support was discussed in relation to visual comparison and data editing tasks. Addressing modern technology, we investigated low-level tracking of tangible views above a tabletop and tracking of users in front of large displays. Finally, we described two approaches to reduce the gulfs of execution and evaluation through automatic methods for the benefit of the human user.

While we considered all aspects individually, we also identified and discussed connections among them. Table 4.1 sets the individual methods and techniques from this chapter in relation to our interaction-oriented view on visualization. The table particularly focuses on the interesting interplay of data, tasks, technology, and human. The implementation aspect is not included in the table. From the descriptions and figures given in this chapter it is evident that appropriate implementation is crucial across all aspects.

An overall conclusion and discussion of the topics covered in this work along with an outlook on future work in the context of interaction in visualization will be given in the next chapter.

CHAPTER 5

Conclusion and Future Work

This final chapter will provide a summarizing discussion of interaction in visualization. In the first section, we will give some concluding remarks and derive take-home messages. In the second section, we will indicate directions for future research with a particular focus on the interaction side of visualization.

5.1 CONCLUDING REMARKS

This work started out with the goal to strengthen the interaction side of visualization. We looked at fundamentals of visualization and interaction and we discussed the challenges involved when designing and implementing interaction techniques in the context of visualization. We presented a unified view of interaction around the key aspects: data, tasks, technology, human, and implementation. Aligned with this view, a number of low-level, intermediate-level, and high-level methods and techniques for interactive visualization were described. In the following paragraphs, we will crystallize the greater picture of this work as a whole.

Unified View and Key Aspects   At the core of this work is the discussion of approaches to interaction in visualization around a unified view. While we identified and discussed key aspects of our interaction-oriented view individually, we also highlighted interrelations among them. To pick only one example from Table 4.1 from the previous chapter, let us once more take a look at the EditLens approach from Section 4.3.2. The EditLens is certainly focused on the specific task of editing data,[1] but it also considers the characteristics of the data (graphs with customized layouts), the specifics of the interaction technology (touch interaction), and the needs of the human user (semi-automatic support). In a similar sense, all other approaches discussed in this work indicate how important each of the key aspects is, both individually and as a cornerstone of our unified view.

In fact, a key message is that one has to consider all aspects of interaction in order to be successful when designing and implementing interaction for visualization. Leaving only one of the aspects out of consideration puts us at risk to arrive at inappropriate or unusable solutions.

Interaction and Automatic Means   Another major point discussed in this work is that interaction, although powerful and effective, is not a universal answer to all problems. This points us to

---

[1]Although editing is not a typical visualization task, interactive correction and manipulation of the data to be analyzed are becoming more and more relevant (Kandel et al., 2011).

think more carefully about where interaction is indeed helpful and in what situations the user is better off with less interaction.

This thinking is reflected several times by approaches that combine interaction with automatic means. Examples were given in Section 4.3.2, where fully manual data editing is facilitated by automatic computations. In Section 4.5.1, automatic event-triggered adjustments of visual representation were applied to reduce interaction costs. Automatic DOI-based concepts are the backbone of the navigation recommendations presented in Section 4.5.2. A less obvious example is the elevator interaction shortcut from Section 4.2.2, which takes the user automatically to a desired position in the visualization space.

The integration of interactive and automatic means has been successful in the mentioned examples. Yet, balancing interaction and automatism is not trivial. Microsoft's famous office assistance Clippit (aka. Clippy) is an example where the balance was off. We would want to avoid intrusive automatic methods at all costs.

Broad Utilization of Technology   The interaction techniques we discussed utilize different technologies. Looking at the display and interaction equipment that we employed to drive the visual interface between the human and the computer, we see a gradual shift from classic desktop settings with mouse and keyboard to modern display technologies and interaction modalities.

Predominant in Sections 4.2.1–4.3.1 was the classic mouse and keyboard setup. But still the interactive approaches presented there are useful and efficient, a fact that indicates that even in standard settings there is still potential to be exploited. In Section 4.3.2, we started to go beyond mouse and keyboard by designing interaction for touch-enabled devices. Advancing further, Section 4.4.1 discussed tangible interaction in the space above a tabletop display. Finally, we also explored proxemic interaction through physical navigation and head tracking in front of a large display wall in Section 4.4.2.

As a conclusion, the approaches presented here suggest that a broad utilization of display and interaction technologies bears much potential for visualization applications. This echoes the statements made by other researchers who argue for taking advantage of modern technology (Lee et al., 2012; Isenberg et al., 2013b; Jakobsen et al., 2013; Keefe and Isenberg, 2013).

Lenses as Universal Tools   A recurring pattern throughout this work is the use of interactive lenses to facilitate visualization tasks. Thanks to the lightweight and focused nature of lenses, they are particularly suited for exploratory visualization.

In Sections 4.2.1 and 4.2.2, we used lenses to support the exploration of graphs and to integrate temporal information into a visual representation that is otherwise focused on spatial aspects. In Section 4.4.2, the user's viewing direction was tracked in order to project a focus+context lens onto a large display wall. While these examples can be considered virtual lenses, which represent a digitized copy of the lens metaphor on the display, the tangible views from Section 4.4.1 are examples of physical, graspable lenses. Such tangible lenses open up whole new possibilities for visualization purposes as has been demonstrated in several case studies.

A look at the literature confirms that many more interactive lenses exist for various types of data and tasks (Tominski et al., 2014). From the wide application of lenses in this work and in the work of others, we can conjecture that lenses are universally useful tools in interactive visualization.

The Challenge of Developing Interaction   This work also raised questions with respect to the implementation of interactive visualization. In Section 4.1, we presented an architecture that can help visualization developers circumvent typical multi-threading pitfalls. But still, developing interactive visualization solutions remains difficult due to the multitude and heterogeneity of influencing factors. Spence (2007) explains this quite illustratively:

> Many ingredients to support representation, presentation and interaction are described [...]: like a good chef or skilled painter, interaction designers must select appropriate ingredients from those available and use established concepts to blend them into a pleasing and effective product. And, as with cooking and painting, good interaction design can be achieved only with practice and the experience of both good and not so good results.
>
> — Spence (2007)

Spence's words already indicate that developing interaction, not only in visualization, is hard. During the development of the approaches presented here, many hours if not weeks have been spent on trying out (which essentially means designing and implementing) and throwing away various interaction solutions until eventually the final result surfaced. Although the costs involved were significant, it is worth investing in interaction, because the benefit for the user is a more productive solution and smooth working experience.

5.2   TOPICS FOR FUTURE WORK

Although we presented a number of interaction techniques they are but pieces in the larger puzzle of the *science of interaction* in visualization research (Pike et al., 2009). This puzzle holds many more unsolved questions that need to be addressed. Where appropriate, we already indicated open issues pertaining to the individual approaches discussed in this work. While it is important to address these issues in the short run, the goal of this section is to provide an agenda of topics for more long-term future research with a focus on interaction in visualization.

Support the Engineering of Interaction   Interaction engineering in the context of visualization remains a complex endeavor. A key difficulty is the dependency of the interaction design on the visualization design and the data. Because the visual representation of the data serves to a large degree as the user interface, interaction can only be implemented efficiently with respect to a good visualization design that has been tested with a reasonable number of datasets. In fact, the problem with this dependency is that a little change in the data and their representation can break the interaction design.

Therefore, new conceptual models and strategies have to be developed to support the human engineer in experimenting effortlessly with different visualization and interaction designs. To this end, the aforementioned dependency has to be relaxed. The benefit of such a relaxation will be a greater flexibility and reduced costs during development.

In addition to such conceptual considerations, there is room for improvements on the practical side of interaction engineering. New toolkits focusing on interaction for visualization are helpful in this regard. A recent example indicating the potential benefits is VisDock, a JavaScript mixin library for cross-cutting interaction (Choi et al., 2015). Moreover, methods should be investigated to supersede the cumbersome and error-prone way of implementing interaction via event handlers in some programming language. Modeling interaction via state machines (Appert and Beaudouin-Lafon, 2008) or declarative interaction design (Satyanarayan et al., 2014) should be incorporated more widely in the context of visualization. Ideally, in the future, interaction for visualization will be modeled using a graphical editor, rather than coded in a programming language.

Go Beyond Mouse and Keyboard   Although mouse and keyboard are still the predominant interaction devices and the regular desktop display is still the most frequently used output device, the future will see a shift toward alternative settings. Direct touch on high-resolution surfaces will most likely replace the standard workplace in the near future. Specialized applications will run in large display environments that track their users and integrate the devices the users bring with them.

While the visualization community has begun to recognize this shift (Lee et al., 2012), still more research is necessary to comprehensively integrate modern display and interaction technologies with visualization approaches. Such research has to tackle several challenges. First, technical issues have to be addressed to create a basis upon which interactive visualization solutions can be built. Second, investigations are needed to determine how and where new technologies can be employed most usefully. Third, studies have to be conducted to evaluate the benefits of the new technologies in comparison to established setups.

A concrete example could be a comparative study on graph exploration. Such a study could compare the classic desktop-based techniques from Section 4.2.1, the tangible views from Section 4.4.1, and the proxemic interaction as described in Section 4.4.2. One can easily imagine how complex such a study would be. On the other hand, such a study would significantly advance our understanding of interaction in visualization.

Provide Guidelines and Guidance   Approaches to interaction in visualization are as multi-faceted as the problems they address. With our interaction-oriented view, we can somewhat structure the space of interaction solutions. Yet, with interaction for different data and different tasks, and maybe even for different technologies, it can become difficult to develop or choose an appropriate interaction technique for a given visualization problem.

Therefore, we have to provide guidelines for visualization engineers and guidance for visualization users. By guidelines we mean a set of established best practices that a visualization engineer can refer to when developing interactive visualization approaches. Such guidelines could, for example, suggest how to map interaction intents to appropriate interaction techniques using the most efficient technology.

While guidelines apply in the development phase of visualization, guidance is to support the user while working with interactive visualization tools. The navigation recommendations from Section 4.5.2 are an example for guidance during interactive navigation. But guidance in general is much broader and can be provided with respect to various domains at different degrees. Here we see much potential for future research on guiding the user in making the most of interactive visualization (Schulz et al., 2013b).

Overcome the Interaction-Visualization Gap   The gap between visualization and interaction has to be narrowed further. A still problematic concern is the lack of a consistent model that covers visual aspects as well as aspects of interaction on equal terms. As a model for interaction, Norman's action cycle describes how humans interact. The visualization pipeline is a model for generating visual representations. In the former case, the human acts upon the data (in visual form), whereas in the latter case, the computer acts upon the data (in symbolic form). These two views have to be brought closer together.

A new integrated model of a visualization-interaction "pipe-loop" would help bridge the gap between interaction and visualization research. One option to build such a model is to compile a conglomerate of the visualization pipeline (Haber and McNabb, 1990; Card et al., 1999; dos Santos and Brodlie, 2004), Norman's (2013) interaction model, and the model of visual exploration by Jankun-Kelly et al. (2007). Constructing such an integrated model requires a detailed analysis of where visualization and interaction models differ and what the key influencing factors behind the models are.

Establish an Interaction Vocabulary   With Bertin's visual variables there is an established vocabulary of basic building blocks for the graphics design of visualization approaches. However, there are no such building blocks for interaction in visualization.

Therefore, an open research question is to define an interaction vocabulary. There have already been first efforts to identify patterns or primitives of interaction in visualization (Sedig and Parsons, 2013; Roth, 2013). These studies describe how interaction is employed for different tasks. What we are still missing are constructive building blocks that allow us to flexibly outfit a visualization design with a suitable interaction design depending on the characteristics of the underlying problem.

Identifying, collecting, and structuring such building blocks in an interaction vocabulary, will enable us to investigate whole new topics. As one such topic we envision adaptive interaction, where interaction techniques automatically adapt to the data and the task at hand as well as to the available technology and the human user operating it. Maybe it is even possible to extend the interaction vocabulary to a *grammar of interaction* analogous to Wilkinson's (2005) *grammar of graphics*.

As mentioned earlier, our ideas for future work represent larger research topics to be studied in the long run. The versatility of the topics suggests that there is still much to do in the context of a *science of interaction* in visualization. We believe that there is much potential in further strengthening the interaction side of visualization. The positive results obtained with the approaches presented in this work let us conclude that it is worth facing and tackling the challenges ahead of us.

Bibliography

J. Abello, S. Hadlak, H. Schumann, and H.-J. Schulz. A Modular Degree-of-Interest Specification for the Visual Analysis of Large Dynamic Networks. *IEEE Transactions on Visualization and Computer Graphics*, 20(3):337–350, 2014. DOI: 10.1109/TVCG.2013.109. 65

C. Ahlberg, C. Williamson, and B. Shneiderman. Dynamic Queries for Information Exploration: An Implementation and Evaluation. In *Proceedings of the SIGCHI Conference Human Factors in Computing Systems (CHI)*, pages 619–626. ACM, 1992. DOI: 10.1145/142750.143054. 11, 15

W. Aigner, S. Miksch, H. Schumann, and C. Tominski. *Visualization of Time-Oriented Data*. Springer, 2011. DOI: 10.1007/978-0-85729-079-3. 40

J. F. Allen, C. I. Guinn, and E. Horvitz. Mixed-Initiative Interaction. *IEEE Intelligent Systems and their Applications*, 14(5):14–23, 1999. DOI: 10.1109/5254.796083. 62

R. Amar, J. Eagan, and J. Stasko. Low-Level Components of Analytic Activity in Information Visualization. In *Proceedings of the IEEE Symposium Information Visualization (InfoVis)*, pages 89–100. IEEE Computer Society, 2005. DOI: 10.1109/INFOVIS.2005.24. 27

F. Amini, S. Rufiange, Z. Hossain, Q. Ventura, P. Irani, and M. J. McGuffin. The Impact of Interactivity on Comprehending 2D and 3D Visualizations of Movement Data. *IEEE Transactions on Visualization and Computer Graphics*, 21(1):122–135, 2015. DOI: 10.1109/TVCG.2014.2329308. 19, 40, 44

C. Andrews and C. North. The Impact of Physical Navigation on Spatial Organization for Sensemaking. *IEEE Transactions on Visualization and Computer Graphics*, 19(12):2207–2216, 2013. DOI: 10.1109/TVCG.2013.205. 58

C. Andrews, A. Endert, B. Yost, and C. North. Information Visualization on Large, High-Resolution Displays: Issues, Challenges, and Opportunities. *Information Visualization*, 10(4): 341–355, 2011. DOI: 10.1177/1473871611415997. 55

G. Andrienko, N. Andrienko, P. Bak, D. Keim, and S. Wrobel. *Visual Analytics of Movement*. Springer, 2013. DOI: 10.1007/978-3-642-37583-5. 44

G. Andrienko, N. Andrienko, H. Schumann, and C. Tominski. Visualization of Trajectory Attributes in Space–Time Cube and Trajectory Wall. In M. Buchroithner, N. Prechtel, and

D. Burghardt, editors, *Cartography from Pole to Pole*, Lecture Notes in Geoinformation and Cartography, pages 157–163. Springer, 2014. DOI: 10.1007/978-3-642-32618-9_11. 44

N. Andrienko and G. Andrienko. *Exploratory Analysis of Spatial and Temporal Data*. Springer, 2006. DOI: 10.1007/3-540-31190-4. 27, 45

C. Appert and M. Beaudouin-Lafon. SwingStates: Adding State Machines to Java and the Swing Toolkit. *Software: Practice and Experience*, 38(11):1149–1182, 2008. DOI: 10.1002/spe.867. 23, 72

B. Bach, E. Pietriga, and J.-D. Fekete. GraphDiaries: Animated Transitions and Temporal Navigation for Dynamic Networks. *IEEE Transactions on Visualization and Computer Graphics*, 20 (5):740–754, 2014. DOI: 10.1109/TVCG.2013.254. 40

R. Ball, C. North, and D. A. Bowman. Move to Improve: Promoting Physical Navigation to Increase User Performance with Large Displays. In *Proceedings of the SIGCHI Conference Human Factors in Computing Systems (CHI)*, pages 191–200. ACM, 2007. DOI: 10.1145/1240624.1240656. 58

T. Ballendat, N. Marquardt, and S. Greenberg. Proxemic Interaction: Designing for a Proximity and Orientation-Aware Environment. In *Proceedings of the International Conference on Interactive Tabletops and Surfaces (ITS)*, pages 121–130. ACM, 2010. DOI: 10.1145/1936652.1936676. 55

G. D. Battista, P. Eades, R. Tamassia, and I. G. Tollis. *Graph Drawing: Algorithms for the Visualization of Graphs*. Prentice Hall, 1999. 37

T. Baudel. From Information Visualization to Direct Manipulation: Extending a Generic Visualization Framework for the Interactive Editing of Large Datasets. In *Proceedings of the ACM Symposium on User Interface Software and Technology (UIST)*, pages 67–76. ACM, 2006. DOI: 10.1145/1166253.1166265. 28, 48, 51

M. Beaudouin-Lafon. Instrumental Interaction: An Interaction Model for Designing Post-WIMP User Interfaces. In *Proceedings of the SIGCHI Conference Human Factors in Computing Systems (CHI)*, pages 446–453. ACM, 2000. DOI: 10.1145/332040.332473. 9

M. Beaudouin-Lafon. Novel Interaction Techniques for Overlapping Windows. In *Proceedings of the ACM Symposium on User Interface Software and Technology (UIST)*, pages 153–154. ACM, 2001. DOI: 10.1145/502348.502371. 48

R. A. Becker and W. S. Cleveland. Brushing Scatterplots. *Technometrics*, 29(2):127–142, 1987. DOI: 10.2307/1269768. 15

B. B. Bederson. The Promise of Zoomable User Interfaces. *Behaviour & Information Technology*, 30(6):853–866, 2011. DOI: 10.1080/0144929X.2011.586724. 13

B. B. Bederson and J. D. Hollan. Pad++: A Zooming Graphical Interface for Exploring Alternate Interface Physics. In *Proceedings of the ACM Symposium on User Interface Software and Technology (UIST)*, pages 17–26. ACM, 1994. DOI: 10.1145/192426.192435. 13

J. Bertin. *Graphics and Graphic Information-Processing*. de Gruyter, 1981. 1

J. Bertin. *Semiology of Graphics: Diagrams, Networks, Maps*. University of Wisconsin Press, 1983. 6, 26

E. A. Bier. *Snap-Dragging: Interactive Geometric Design in Two and Three Dimensions*. PhD thesis, University of California, Berkeley, 1988. URL http://www.eecs.berkeley.edu/Pubs/TechRpts/1988/5858.html. 47

M. Bostock, V. Ogievetsky, and J. Heer. D^3 Data-Driven Documents. *IEEE Transactions on Visualization and Computer Graphics*, 17(12):2301–2309, 2011. DOI: 10.1109/TVCG.2011.185. 22

F. Bouali, A. Guettala, and G. Venturini. VizAssist: An Interactive User Assistant for Visual Data Mining. *The Visual Computer*, pages 1–17, 2015. DOI: 10.1007/s00371-015-1132-9. 62

A. Buja, J. A. McDonald, J. Michalak, and W. Stuetzle. Interactive Data Visualization using Focusing and Linking. In *Proceedings of the IEEE Visualization Conference (Vis)*, pages 156–163, 419. IEEE Computer Society, 1991. DOI: 10.1109/VISUAL.1991.175794. 15

S. K. Card, G. G. Robertson, and J. D. Mackinlay. The Information Visualizer, an Information Workspace. In *Proceedings of the SIGCHI Conference Human Factors in Computing Systems (CHI)*, pages 181–186. ACM, 1991. DOI: 10.1145/108844.108874. 30

S. K. Card, J. D. Mackinlay, and B. Shneiderman. *Readings in Information Visualization: Using Vision to Think*. Morgan Kaufmann, 1999. 5, 6, 9, 73

S.-M. Chan, L. Xiao, J. Gerth, and P. Hanrahan. Maintaining Interactivity While Exploring Massive Time Series. In *Proceedings of the IEEE Symposium on Visual Analytics Science and Technology (VAST)*, pages 59–66. IEEE Computer Society, 2008. DOI: 10.1109/VAST.2008.4677357. 36

H. Chen. Compound Brushing Explained. *Information Visualization*, 3(2):96–108, 2004. DOI: 10.1057/palgrave.ivs.9500068. 15

E. H. Chi. A Taxonomy of Visualization Techniques Using the Data State Reference Model. In *Proceedings of the IEEE Symposium Information Visualization (InfoVis)*, pages 69–75. IEEE Computer Society, 2000. DOI: 10.1109/INFVIS.2000.885092. 6

L. Chittaro. Visualizing Information on Mobile Devices. *IEEE Computer*, 39(3):40–45, 2006. DOI: 10.1109/MC.2006.109. 28

J. Choi, D. G. Park, Y. L. Wong, E. Fisher, and N. Elmqvist. VisDock: A Toolkit for Cross-Cutting Interactions in Visualization. *IEEE Transactions on Visualization and Computer Graphics*, 2015. DOI: 10.1109/TVCG.2015.2414454. to appear. 72

M. C. Chuah and S. F. Roth. On the semantics of interactive visualizations. In *Proceedings of the IEEE Symposium Information Visualization (InfoVis)*, pages 29–36. IEEE Computer Society, 1996. DOI: 10.1109/INFVIS.1996.559213. 27

W. S. Cleveland and R. McGill. An Experiment in Graphical Perception. *International Journal of Man-Machine Studies*, 25(5):491–501, 1986. DOI: 10.1016/S0020-7373(86)80019-0. 26

D. M. Coffey, N. Malbraaten, T. B. Le, I. Borazjani, F. Sotiropoulos, A. G. Erdman, and D. F. Keefe. Interactive Slice WIM: Navigating and Interrogating Volume Data Sets Using a Multi-surface, Multitouch VR Interface. *IEEE Transactions on Visualization and Computer Graphics*, 18(10):1614–1626, 2012. DOI: 10.1109/TVCG.2011.283. 29

S. Conversy. Improving Usability of Interactive Graphics Specification and Implementation with Picking Views and Inverse Transformation. In *Proceedings of the IEEE Symposium on Visual Languages and Human-Centric Computing (VL/HCC)*, pages 153–160. IEEE Computer Society, 2011. DOI: 10.1109/VLHCC.2011.6070392. 23

S. Conversy. Contributions to the Science of Controlled Transformation. Habilitation à Diriger des Recherches de l'Université Paul Sabatier - Toulouse III, 2013. URL http://tel.arch ives-ouvertes.fr/tel-00853192. 10, 31

A. Cooper, R. Reimann, and D. Cronin. *About Face 3: The Essentials of Interaction Design*. Wiley, 2007. 20, 21, 65

J. A. Cottam, A. Lumsdaine, and P. Wang. Abstract Rendering: Out-of-core Rendering for Information Visualization. In *Proceedings of the Conference on Visualization and Data Analysis (VDA)*, pages 90170K–1–90170K–13. SPIE, 2014. DOI: 10.1117/12.2041200. 36

T. Crnovrsanin, I. Liao, Y. Wu, and K.-L. Ma. Visual Recommendations for Network Navigation. *Computer Graphics Forum*, 30(3):1081–1090, 2011. DOI: 10.1111/j.1467-8659.2011.01957.x. 65

A. Dix, J. Finlay, G. D. Abowd, and R. Beale. *Human-Computer Interaction*. Pearson Education, 3rd edition, 2004. 7, 9

H. Doleisch. SimVis: Interactive Visual Analysis of Large and Time-Dependent 3D Simulation Data. In *Proceedings of the Winter Simulation Conference*, pages 712–720. IEEE Computer Society, 2007. DOI: 10.1109/WSC.2007.4419665. 35

S. dos Santos and K. Brodlie. Gaining Understanding of Multivariate and Multidimensional Data through Visualization. *Computers & Graphics*, 28:311–325, 2004. DOI: 10.1016/j.cag.2004.03.013. 6, 73

P. Dragicevic. Combining Crossing-Based and Paper-Based Interaction Paradigms for Dragging and Dropping Between Overlapping Windows. In *Proceedings of the ACM Symposium on User Interface Software and Technology (UIST)*, pages 193–196. ACM, 2004. DOI: 10.1145/1029632.1029667. 48

T. Dwyer, K. Marriott, F. Schreiber, P. Stuckey, M. Woodward, and M. Wybrow. Exploration of Networks Using Overview+Detail with Constraint-Based Cooperative Layout. *IEEE Transactions on Visualization and Computer Graphics*, 14(6):1293–1300, 2008. DOI: 10.1109/TVCG.2008.130. 51

A. Ebert, G. C. van der Veer, G. Domik, N. D. Gershon, and I. Scheler, editors. *Building Bridges: HCI, Visualization, and Non-formal Modeling*. Springer, 2014. DOI: 10.1007/978-3-642-54894-9. 17, 18

J. Edwards. Coherent Reaction. In *Companion to the 24th Annual ACM SIGPLAN Conference on Object-Oriented Programming, Systems, Languages, and Applications (OOPSLA)*, pages 925–932. ACM, 2009. DOI: 10.1145/1639950.1640058. 22

N. Elmqvist and J.-D. Fekete. Hierarchical Aggregation for Information Visualization: Overview, Techniques, and Design Guidelines. *IEEE Transactions on Visualization and Computer Graphics*, 16(3):439–454, 2010. DOI: 10.1109/TVCG.2009.84. 63

N. Elmqvist, A. V. Moere, H.-C. Jetter, D. Cernea, H. Reiterer, and T. Jankun-Kelly. Fluid Interaction for Information Visualization. *Information Visualization*, 10(4):327–340, 2011. DOI: 10.1177/1473871611413180. 2, 11, 19, 29

J.-D. Fekete. Advanced interaction for Information Visualization. In *Proceedings of the IEEE Pacific Visualization Symposium (PacificVis)*, page xi. IEEE Computer Society, 2010. DOI: 10.1109/PACIFICVIS.2010.5429617. 2

J.-D. Fekete. Visual Analytics Infrastructures: From Data Management to Exploration. *IEEE Computer*, 46(7):22–29, 2013. DOI: 10.1109/MC.2013.120. 31

J.-D. Fekete and C. Plaisant. Interactive Information Visualization of a Million Items. In *Proceedings of the IEEE Symposium Information Visualization (InfoVis)*, pages 117–124. IEEE Computer Society, 2002. DOI: 10.1109/INFVIS.2002.1173156. 36

J.-D. Fekete, P.-L. Hemery, T. Baudel, and J. Wood. Obvious: A Meta-Toolkit to Encapsulate Information Visualization Toolkits - One Toolkit to Bind Them All. In *Proceedings of the IEEE Symposium on Visual Analytics Science and Technology (VAST)*, pages 91–100. IEEE Computer Society, 2011. DOI: 10.1109/VAST.2011.6102446. 22

W. Fikkert, M. D'Ambros, T. Bierz, and T. Jankun-Kelly. Interacting with Visualizations. In A. Kerren, A. Ebert, and J. Meyer, editors, *Human-Centered Visualization Environments*, volume 4417 of *Lecture Notes in Computer Science*, pages 77–162. Springer, 2007. DOI: 10.1007/978-3-540-71949-6_3. 18

C. Forlines and R. H. Lilien. Adapting a Single-User, Single-Display Molecular Visualization Application for Use in a Multi-User, Multi-Display Environment. In *Proceedings of the Conference on Advanced Visual Interfaces (AVI)*, pages 367–371. ACM, 2008. DOI: 10.1145/1385569.1385635. 28

M. Frisch. *Interaction and Visualization Techniques for Node-Link Diagram Editing and Exploration*. PhD thesis, Otto-von-Guericke-Universität Magdeburg, 2012. 50

M. Frisch and R. Dachselt. Visualizing Offscreen Elements of Node-Link Diagrams. *Information Visualization*, 12(2):133–162, 2013. DOI: 10.1177/1473871612473589. 40

M. Frisch, J. Heydekorn, and R. Dachselt. Investigating Multi-Touch and Pen Gestures for Diagram Editing on Interactive Surfaces. In *Proceedings of the International Conference on Interactive Tabletops and Surfaces (ITS)*, pages 149–156. ACM, 2009. DOI: 10.1145/1731903.1731933. 51

G. W. Furnas. Generalized Fisheye Views. *ACM SIGCHI Bulletin*, 17(4):16–23, 1986. DOI: 10.1145/22339.22342. 65

G. W. Furnas. Effective View Navigation. In *Proceedings of the SIGCHI Conference Human Factors in Computing Systems (CHI)*, pages 367–374. ACM, 1997. DOI: 10.1145/258549.258800. 62

G. W. Furnas and B. B. Bederson. Space-Scale Diagrams: Understanding Multiscale Interfaces. In *Proceedings of the SIGCHI Conference Human Factors in Computing Systems (CHI)*, pages 234–241. ACM, 1995. DOI: 10.1145/223904.223934. 13

S. Ghani, N. H. Riche, and N. Elmqvist. Dynamic Insets for Context-Aware Graph Navigation. *Computer Graphics Forum*, 30(3):861–870, 2011a. DOI: 10.1111/j.1467-8659.2011.01935.x. 40

S. Ghani, N. H. Riche, and N. Elmqvist. Dynamic Insets for Context-Aware Graph Navigation. *Computer Graphics Forum*, 30(3):861–870, 2011b. DOI: 10.1111/j.1467-8659.2011.01935.x. 40

S. Gladisch, H. Schumann, and C. Tominski. Navigation Recommendations for Exploring Hierarchical Graphs. In G. Bebis, R. Boyle, B. Parvin, D. Koracin, B. Li, F. Porikli, V. Zordan, J. Klosowski, S. Coquillart, X. Luo, M. Chen, and D. Gotz, editors, *Advances in Visual Computing*, pages 36–47. Springer, 2013. DOI: 10.1007/978-3-642-41939-3_4. 63, 65

S. Gladisch, H. Schumann, M. Ernst, G. Füllen, and C. Tominski. Semi-Automatic Editing of Graphs with Customized Layouts. *Computer Graphics Forum*, 33(3):381–390, 2014. DOI: 10.1111/cgf.12394. 48, 49, 50

M. Gleicher. Image Snapping. In *Proceedings of the Annual Conference on Computer Graphics and Interactive Techniques (SIGGRAPH)*, pages 183–190. ACM, 1995. DOI: 10.1145/218380.218441. 47

M. Gleicher, D. Albers, R. Walker, I. Jusufi, C. D. Hansen, and J. C. Roberts. Visual Comparison for Information Visualization. *Information Visualization*, 10(4):289–309, 2011. DOI: 10.1177/1473871611416549. 45, 48

D. Gotz and Z. Wen. Behavior-driven Visualization Recommendation. In *Proceedings of the International Conference on Intelligent User Interfaces (IUI)*, pages 315–324. ACM, 2009. DOI: 10.1145/1502650.1502695. 62

S. Greenberg, S. Boring, J. Vermeulen, and J. Dostal. Dark Patterns in Proxemic Interactions: A Critical Perspective. In *Proceedings of theDesigning Interactive Systems Conference (DIS)*, pages 523–532. ACM, 2014. DOI: 10.1145/2598510.2598541. 58

E. Grundy, M. W. Jones, R. S. Laramee, R. P. Wilson, and E. L. C. Shepard. Visualisation of Sensor Data from Animal Movement. *Computer Graphics Forum*, 28(3):815–822, 2009. DOI: 10.1111/j.1467-8659.2009.01469.x. 40

S. Gustafson, P. Baudisch, C. Gutwin, and P. Irani. Wedge: Clutter-Free Visualization of Off-Screen Locations. In *Proceedings of the SIGCHI Conference Human Factors in Computing Systems (CHI)*, pages 787–796. ACM, 2008. DOI: 10.1145/1357054.1357179. 65

R. B. Haber and D. A. McNabb. Visualization Idioms: A Conceptual Model for Scientific Visualization Systems. In G. M. Nielson, B. D. Shriver, and L. J. Rosenblum, editors, *Visualization in Scientific Computing*, pages 74–93. IEEE Computer Society, 1990. 6, 73

T. Hakala, J. Lehikoinen, and A. Aaltonen. Spatial Interactive Visualization on Small Screen. In *Proceedings of the International Conference on Human Computer Interaction with Mobile Devices & Services (MobileCHI)*, pages 137–144. ACM, 2005. DOI: 10.1145/1085777.1085800. 28

M. Hascoët and P. Dragicevic. Interactive Graph Matching and Visual Comparison of Graphs and Clustered Graphs. In *Proceedings of the Conference on Advanced Visual Interfaces (AVI)*, pages 522–529. ACM, 2012. DOI: 10.1145/2254556.2254654. 40

M. Hassenzahl and N. Tractinsky. User Experience – A Research Agenda. *Behaviour Information Technology*, 25(2):91–97, 2006. DOI: 10.1080/01449290500330331. 8, 29

H. Hauser, F. Ledermann, and H. Doleisch. Angular Brushing of Extended Parallel Coordinates. In *Proceedings of the IEEE Symposium Information Visualization (InfoVis)*, pages 127–130. IEEE Computer Society, 2002. DOI: 10.1109/INFVIS.2002.1173157. 15

C. G. Healey, S. Kocherlakota, V. Rao, R. Mehta, and R. St.Amant. Visual Perception and Mixed-Initiative Interaction for Assisted Visualization Design. *IEEE Transactions on Visualization and Computer Graphics*, 14(2):396–411, 2008. DOI: 10.1109/TVCG.2007.70436. 62

J. Heer and M. Agrawala. Software Design Patterns for Information Visualization. *IEEE Transactions on Visualization and Computer Graphics*, 12(5):853–860, 2006. DOI: 10.1109/TVCG.2006.178. 22

J. Heer and M. Bostock. Crowdsourcing Graphical Perception: Using Mechanical Turk to Assess Visualization Design. In *Proceedings of the SIGCHI Conference Human Factors in Computing Systems (CHI)*, pages 203–212. ACM, 2010. DOI: 10.1145/1753326.1753357. 6, 26

J. Heer and G. Robertson. Animated Transitions in Statistical Data Graphics. *IEEE Transactions on Visualization and Computer Graphics*, 13(6):1240–1247, 2007. DOI: 10.1109/TVCG.2007.70539. 31

J. Heer and B. Shneiderman. Interactive Dynamics for Visual Analysis. *Communications of the ACM*, 55(4):45–54, 2012. DOI: 10.1145/2133806.2133821. 11, 34

K. Henriksen, J. Sporring, and K. Hornbæk. Virtual Trackballs Revisited. *IEEE Transactions on Visualization and Computer Graphics*, 10(2):206–216, 2004. DOI: 10.1109/TVCG.2004.1260772. 41

H. Hochheiser and B. Shneiderman. Dynamic Query Tools for Time Series Data Sets: Timebox Widgets for Interactive Exploration. *Information Visualization*, 3(1):1–18, 2004. DOI: 10.1057/palgrave.ivs.9500061. 27

D. Holman, R. Vertegaal, M. Altosaar, N. F. Troje, and D. Johns. Paper Windows: Interaction Techniques for Digital Paper. In *Proceedings of the SIGCHI Conference Human Factors in Computing Systems (CHI)*, pages 591–599. ACM, 2005. DOI: 10.1145/1054972.1055054. 55

C. Holz and S. Feiner. Relaxed Selection Techniques for Querying Time-Series Graphs. In *Proceedings of the ACM Symposium on User Interface Software and Technology (UIST)*, pages 213–222. ACM, 2009. DOI: 10.1145/1622176.1622217. 27

K. Hornbæk, B. B. Bederson, and C. Plaisant. Navigation Patterns and Usability of Zoomable User Interfaces with and without an Overview. *ACM Transactions on Computer-Human Interaction*, 9(4):362–389, 2002. DOI: 10.1145/586081.586086. 14

E. Horvitz. Principles of Mixed-Initiative User Interfaces. In *Proceedings of the SIGCHI Conference Human Factors in Computing Systems (CHI)*, pages 159–166. ACM, 1999. DOI: 10.1145/302979.303030. 62

W. Huang, editor. *Handbook of Human Centric Visualization*. Springer, 2013. DOI: 10.1007/978-1-4614-7485-2. 29

C. Hurter, B. Tissoires, and S. Conversy. FromDaDy: Spreading Aircraft Trajectories Across Views to Support Iterative Queries. *IEEE Transactions on Visualization and Computer Graphics*, 15(6):1017–1024, 2009. DOI: 10.1109/TVCG.2009.145. 40

J.-F. Im, F. G. Villegas, and M. J. McGuffin. VisReduce: Fast and Responsive Incremental Information Visualization of Large Datasets. In *Proceedings of the IEEE International Conference on Big Data*, pages 25–32. IEEE Computer Society, 2013. DOI: 10.1109/BigData.2013.6691710. 36

P. Isenberg and M. S. T. Carpendale. Interactive Tree Comparison for Co-located Collaborative Information Visualization. *IEEE Transactions on Visualization and Computer Graphics*, 13(6): 1232–1239, 2007. DOI: 10.1109/TVCG.2007.70568. 48

P. Isenberg, P. Dragicevic, W. Willett, A. Bezerianos, and J.-D. Fekete. Hybrid-Image Visualization for Large Viewing Environments. *IEEE Transactions on Visualization and Computer Graphics*, 19(12):2346–2355, 2013a. DOI: 10.1109/TVCG.2013.163. 58

P. Isenberg, T. Isenberg, T. Hesselmann, B. Lee, U. von Zadow, and A. Tang. Data Visualization on Interactive Surfaces: A Research Agenda. *Computer Graphics and Applications*, 33(2):16–24, 2013b. DOI: 10.1109/MCG.2013.24. 19, 29, 58, 70

H. Ishii and B. Ullmer. Tangible Bits: Towards Seamless Interfaces Between People, Bits and Atoms. In *Proceedings of the SIGCHI Conference Human Factors in Computing Systems (CHI)*, pages 234–241. ACM, 1997. DOI: 10.1145/258549.258715. 9, 55

B. Jackson, T. Y. Lau, D. Schroeder, K. C. Toussaint, and D. F. Keefe. A Lightweight Tangible 3D Interface for Interactive Visualization of Thin Fiber Structures. *IEEE Transactions on Visualization and Computer Graphics*, 19(12):2802–2809, 2013. DOI: 10.1109/TVCG.2013.121. 55

R. J. Jacob, A. Girouard, L. M. Hirshfield, M. S. Horn, O. Shaer, E. T. Solovey, and J. Zigelbaum. Reality-Based Interaction: A Framework for Post-WIMP Interfaces. In *Proceedings of the SIGCHI Conference Human Factors in Computing Systems (CHI)*, pages 201–210. ACM, 2008. DOI: 10.1145/1357054.1357089. 9, 30

M. R. Jakobsen and K. Hornbæk. Is Moving Improving?: Some Effects of Locomotion in Wall-Display Interaction. In *Proceedings of the SIGCHI Conference Human Factors in Computing Systems (CHI)*, pages 4169–4178. ACM, 2015. DOI: 10.1145/2702123.2702312. 58

M. R. Jakobsen, Y. S. Haile, S. Knudsen, and K. Hornbæk. Information Visualization and Proxemics: Design Opportunities and Empirical Findings. *IEEE Transactions on Visualization and
Computer Graphics*, 19(12):2386–2395, 2013. DOI: 10.1109/TVCG.2013.166. 58, 70

J. Jankowski and M. Hachet. A Survey of Interaction Techniques for Interactive 3D Environments. In *Eurographics - State of the Art Reports*. Eurographics Association, 2013. DOI:
10.2312/conf/EG2013/stars/065-093. 17, 41, 44

T. Jankun-Kelly, K.-L. Ma, and M. Gertz. A Model and Framework for Visualization Exploration. *IEEE Transactions on Visualization and Computer Graphics*, 13(2):357–369, 2007. DOI:
10.1109/TVCG.2007.28. 22, 35, 73

Y. Jansen and P. Dragicevic. An Interaction Model for Visualizations Beyond The Desktop.
IEEE Transactions on Visualization and Computer Graphics, 19(12):2396–2405, 2013. DOI:
10.1109/TVCG.2013.134. 6, 58

S. Kandel, J. Heer, C. Plaisant, J. Kennedy, F. van Ham, N. H. Riche, C. Weaver, B. Lee,
D. Brodbeck, and P. Buono. Research Directions in Data Wrangling: Visualizations and
Transformations for Usable and Credible Data. *Information Visualization*, 10(4):271–288,
2011. DOI: 10.1177/1473871611415994. 28, 48, 51, 69

D. F. Keefe. Integrating Visualization and Interaction Research to Improve Scientific Workflows.
Computer Graphics and Applications, 30(2):8–13, 2010. DOI: 10.1109/MCG.2010.30. 2, 18

D. F. Keefe and T. Isenberg. Reimagining the Scientific Visualization Interaction Paradigm.
IEEE Computer, 46(5):51–57, 2013. DOI: 10.1109/MC.2013.178. 29, 30, 58, 70

D. F. Keefe, D. Acevedo, J. Miles, F. Drury, S. Swartz, and D. Laidlaw. Scientific Sketching
for Collaborative VR Visualization Design. *IEEE Transactions on Visualization and Computer
Graphics*, 14(4):835–847, 2008. DOI: 10.1109/TVCG.2008.31. 29

D. F. Keefe, A. Gupta, D. Feldman, J. V. Carlis, S. K. Keefe, and T. J. Griffin. Scaling Up
Multi-Touch Selection and Querying: Interfaces and Applications for Combining Mobile
Multi-Touch Input with Large-Scale Visualization Displays. *International Journal of Human-
Computer Studies*, 70(10):703–713, 2012. DOI: 10.1016/j.ijhcs.2012.05.004. 58

D. Keim, J. Kohlhammer, G. Ellis, and F. Mansmann, editors. *Mastering the Information Age –
Solving Problems with Visual Analytics*. Eurographics Association, 2010. 31

D. A. Keim, F. Mansmann, D. Oelke, and H. Ziegler. Visual Analytics: Combining Automated
Discovery with Interactive Visualizations. In J. Boulicaut, M. R. Berthold, and T. Horváth, editors, *Discovery Science*, volume 5255 of *Lecture Notes in Computer Science*, pages 2–14. Springer,
2008. DOI: 10.1007/978-3-540-88411-8_2. 62

A. Kerren, A. Ebert, and J. Meyer, editors. *Human-Centered Visualization Environments.* Springer, 2007. DOI: 10.1007/978-3-540-71949-6. 29

K. Kim and N. Elmqvist. Embodied lenses for collaborative visual queries on tabletop displays. *Information Visualization*, 11(4):319–338, 2012. DOI: 10.1177/1473871612441874. 52, 55

K. Kin, B. Hartmann, T. DeRose, and M. Agrawala. Proton: Multitouch Gestures as Regular Expressions. In *Proceedings of the SIGCHI Conference Human Factors in Computing Systems (CHI)*, pages 2885–2894. ACM, 2012. DOI: 10.1145/2207676.2208694. 23

Kitware, Inc. *The VTK User's Guide.* Kitware, Inc., 2010. 22

T. R. Klein, F. Guéniat, L. Pastur, F. Vernier, and T. Isenberg. A Design Study of Direct-Touch Interaction for Exploratory 3D Scientific Visualization. *Computer Graphics Forum*, 31(3):1225–1234, 2012. DOI: 10.1111/j.1467-8659.2012.03115.x. 29

P. Kohlmann, S. Bruckner, A. Kanitsar, and M. E. Gröller. Contextual Picking of Volumetric Structures. In *Proceedings of the IEEE Pacific Visualization Symposium (PacificVis)*, pages 185–192. IEEE Computer Society, 2009. DOI: 10.1109/PACIFICVIS.2009.4906855. 17

R. Kosara. Indirect Multi-Touch Interaction for Brushing in Parallel Coordinates. In *Proceedings of the Conference on Visualization and Data Analysis (VDA)*, pages 786809–1–786809–7. SPIE, 2011. DOI: 10.1117/12.872645. 29, 52

M.-J. Kraak. The Space-Time Cube Revisited from a Geovisualization Perspective. In *Proceedings of the International Cartographic Conference (ICC)*, pages 1988–1995. The International Cartographic Association (ICA), 2003. 61

M.-J. Kraak and F. J. Ormeling. *Cartography: Visualization of Spatial Data.* Pearson Education, 2010. 40

G. E. Krasner and S. T. Pope. A Cookbook for Using the Model-View-Controller User Interface Paradigm in Smalltalk-80. *Journal of Object-Oriented Programming*, 1(3):26–49, 1988. 22

M. Kreuseler and H. Schumann. A Flexible Approach for Visual Data Mining. *IEEE Transactions on Visualization and Computer Graphics*, 8(1):39–51, 2002. DOI: 10.1109/2945.981850. 36

M. Kreuseler, T. Nocke, and H. Schumann. A History Mechanism for Visual Data Mining. In *Proceedings of the IEEE Symposium Information Visualization (InfoVis)*, pages 49–56. IEEE Computer Society, 2004. DOI: 10.1109/INFVIS.2004.2. 22

R. Krüger, D. Thom, M. Wörner, H. Bosch, and T. Ertl. TrajectoryLenses – A Set-based Filtering and Exploration Technique for Long-term Trajectory Data. *Computer Graphics Forum*, 32(3): 451–460, 2013. DOI: 10.1111/cgf.12132. 40, 44

H. Lam.   A Framework of Interaction Costs in Information Visualization.   *IEEE Transactions on Visualization and Computer Graphics*, 14(6):1149–1156, 2008.   DOI: 10.1109/TVCG.2008.109. 21, 29, 59

C. Law, W. Schroeder, K. Martin, and J. Temkin. A Multi-Threaded Streaming Pipeline Architecture for Large Structured Data Sets. In *Proceedings of the IEEE Visualization Conference (Vis)*, pages 225–232. IEEE Computer Society, 1999. DOI: 10.1109/VISUAL.1999.809891. 36

B. Lee, C. Plaisant, C. S. Parr, J.-D. Fekete, and N. Henry. Task Taxonomy for Graph Visualization. In *Proceedings of the Workshop on Beyond Time and Errors: Novel Evaluation Methods for Information Visualization (BELIV)*, pages 1–5. ACM, 2006. DOI: 10.1145/1168149.1168168. 27, 40

B. Lee, P. Isenberg, N. Riche, and S. Carpendale. Beyond Mouse and Keyboard: Expanding Design Considerations for Information Visualization Interactions. *IEEE Transactions on Visualization and Computer Graphics*, 18(12):2689–2698, 2012. DOI: 10.1109/TVCG.2012.204. 19, 29, 70, 72

E. A. Lee.   The Problem with Threads.   *IEEE Computer*, 39(5):33–42, 2006.   DOI: 10.1109/MC.2006.180. 34

A. Lehmann, H. Schumann, O. Staadt, and C. Tominski. Physical Navigation to Support Graph Exploration on a Large High-Resolution Display. In G. Bebis, R. Boyle, B. Parvin, D. Koracin, S. Wang, K. Kyungnam, B. Benes, K. Moreland, C. Borst, S. DiVerdi, C. Yi-Jen, and J. Ming, editors, *Advances in Visual Computing*, volume 6938 of *Lecture Notes in Computer Science*, pages 496–507. Springer, 2011. DOI: 10.1007/978-3-642-24028-7_46. 55, 57

C. Letondal, S. Chatty, W. G. Phillips, F. André, and S. Conversy. Usability Requirements for Interaction-Oriented Development Tools. In *Proceedings of the 22nd Annual Workshop of the Psychology of Programming Interest Group (PPIG)*, pages 12–26, 2010. URL http://www.ppig.org/papers/22nd-UX-2.pdf. 31

L. D. Lins, J. T. Klosowski, and C. E. Scheidegger. Nanocubes for Real-Time Exploration of Spatiotemporal Datasets. *IEEE Transactions on Visualization and Computer Graphics*, 19(12): 2456–2465, 2013. DOI: 10.1109/TVCG.2013.179. 36

Z. Liu and J. Heer.   The Effects of Interactive Latency on Exploratory Visual Analysis.   *IEEE Transactions on Visualization and Computer Graphics*, 20(12):2122–2131, 2014. DOI: 10.1109/TVCG.2014.2346452. 30

Z. Liu and J. Stasko. Mental Models, Visual Reasoning and Interaction in Information Visualization: A Top-down Perspective. *IEEE Transactions on Visualization and Computer Graphics*, 16(6):999–1008, 2010. DOI: 10.1109/TVCG.2010.177. 12

Z. Liu, N. Nersessian, and J. Stasko. Distributed Cognition as a Theoretical Framework for Information Visualization. *IEEE Transactions on Visualization and Computer Graphics*, 14(6): 1173–1180, 2008. DOI: 10.1109/TVCG.2008.121. 12

Z. Liu, B. Jiang, and J. Heer. *imMens*: Real-time Visual Querying of Big Data. *Computer Graphics Forum*, 32(3):421–430, 2013. DOI: 10.1111/cgf.12129. 36

A. M. MacEachren. *Some Truth With Maps: A Primer on Symbolization and Design*. Association of American Geographers, 1994. 6, 40

J. Mackinlay. Automating the Design of Graphical Presentations of Relational Information. *ACM Transactions on Graphics*, 5(2):110–141, 1986. DOI: 10.1145/22949.22950. 6, 26, 29

T. May, M. Steiger, J. Davey, and J. Kohlhammer. Using Signposts for Navigation in Large Graphs. *Computer Graphics Forum*, 31(3pt2):985–994, 2012. DOI: 10.1111/j.1467-8659.2012.03091.x. 65

B. H. McCormick, T. A. DeFanti, and M. D. Brown. Visualization in Scientific Computing. *ACM SIGRAPH Computer Graphics*, 21(6):3, 1987. DOI: 10.1145/41997.41998. 5

M. J. McGuffin and I. Jurisica. Interaction Techniques for Selecting and Manipulating Subgraphs in Network Visualizations. *IEEE Transactions on Visualization and Computer Graphics*, 15(6): 937–944, 2009. DOI: 10.1109/TVCG.2009.151. 37, 51

K. Moreland. A Survey of Visualization Pipelines. *IEEE Transactions on Visualization and Computer Graphics*, 19(3):367–378, 2013. DOI: 10.1109/TVCG.2012.133. 22, 36

T. Moscovich, F. Chevalier, N. Henry, E. Pietriga, and J.-D. Fekete. Topology-Aware Navigation in Large Networks. In *Proceedings of the SIGCHI Conference Human Factors in Computing Systems (CHI)*, pages 2319–2328. ACM, 2009. DOI: 10.1145/1518701.1519056. 40

T. Möller. What is Visualization? Capstone talk at EuroVis, 2012. https://www.cs.sfu.ca/~torsten/Publications/eurovis_120608.pdf, Retrieved: June, 2015.

T. Mühlbacher, H. Piringer, S. Gratzl, M. Sedlmair, and M. Streit. Opening the Black Box: Strategies for Increased User Involvement in Existing Algorithm Implementations. *IEEE Transactions on Visualization and Computer Graphics*, 20(12):1643–1652, 2014. DOI: 10.1109/TVCG.2014.2346578. 36

J. Nielsen. *Usability Engineering*. Morgan Kaufmann, 1993. 8, 29

A. Nocaj and U. Brandes. Computing Voronoi Treemaps: Faster, Simpler, and Resolution-independent. *Computer Graphics Forum*, 31(3):855–864, 2012. DOI: 10.1111/j.1467-8659.2012.03078.x. 30

D. A. Norman. *The Psychology of Everyday Things*. Basic Books, 1988. 7

D. A. Norman. *The Design of Everyday Things*. Basic Books, revised and expanded edition, 2013. 7, 8, 20, 29, 59, 73

M. Okoe, S. S. Alam, and R. Jianu. A Gaze-enabled Graph Visualization to Improve Graph Reading Tasks. *Computer Graphics Forum*, 33(3):251–260, 2014. DOI: 10.1111/cgf.12381. 65

D. R. Olsen. *Building Interactive Systems: Principles for Human-Computer Interaction*. Course Technology, 2009. 31

W. A. Pike, J. T. Stasko, R. Chang, and T. A. O'Connell. The Science of Interaction. *Information Visualization*, 8(4):263–274, 2009. DOI: 10.1057/ivs.2009.22. 1, 18, 71

H. Piringer, W. Berger, and H. Hauser. Quantifying and Comparing Features in High-Dimensional Datasets. In *Proceedings of the International Conference Information Visualisation (IV)*, pages 240–245. IEEE Computer Society, 2008. DOI: 10.1109/IV.2008.17. 35

H. Piringer, C. Tominski, P. Muigg, and W. Berger. A Multi-Threading Architecture to Support Interactive Visual Exploration. *IEEE Transactions on Visualization and Computer Graphics*, 15 (6):1113–1120, 2009. DOI: 10.1109/TVCG.2009.110. 34, 35

P. Pirolli and S. Card. The Sensemaking Process and Leverage Points for Analyst Technology as Identified Through Cognitive Task Analysis. In *Proceedings of the International Conference on Intelligence Analysis*, 2005. 12

Z. Pousman, J. T. Stasko, and M. Mateas. Casual Information Visualization: Depictions of Data in Everyday Life. *IEEE Transactions on Visualization and Computer Graphics*, 13(6):1145–1152, 2007. DOI: 10.1109/TVCG.2007.70541. 18

K. Pulo. Navani: Navigating Large-Scale Visualisations with Animated Transitions. In *Proceedings of the International Conference Information Visualisation (IV)*, pages 271–276. IEEE Computer Society, 2007. DOI: 10.1109/IV.2007.82. 31

A. Radloff, C. Tominski, T. Nocke, and H. Schumann. Supporting Presentation and Discussion of Visualization Results in Smart Meeting Rooms. *The Visual Computer*, 2015. DOI: 10.1007/s00371-014-1010-x. to appear. 28, 58

R. Rao and S. K. Card. The Table Lens: Merging Graphical and Symbolic Representations in an Interactive Focus + Context Visualization for Tabular Information. In *Proceedings of the SIGCHI Conference Human Factors in Computing Systems (CHI)*, pages 318–322. ACM, 1994. DOI: 10.1145/191666.191776. 61

T.-M. Rhyne, M. Tory, T. Munzner, M. Ward, C. Johnson, and D. H. Laidlaw. Information and Scientific Visualization: Separate but Equal or Happy Together at Last. In *Proceedings of the IEEE Visualization Conference (Vis)*, pages 611–614. IEEE Computer Society, 2003. DOI: 10.1109/VISUAL.2003.1250428. Panel discussion.

N. H. Riche, B. Lee, and C. Plaisant. Understanding Interactive Legends: a Comparative Evaluation with Standard Widgets. *Computer Graphics Forum*, 29(3):1193–1202, 2010a. DOI: 10.1111/j.1467-8659.2009.01678.x. 26

Y. Riche, N. H. Riche, P. Isenberg, and A. Bezerianos. Hard-To-Use Interfaces Considered Beneficial (Some of the Time). In *CHI Extended Abstracts on Human Factors in Computing Systems*, pages 2705–2714. ACM, 2010b. DOI: 10.1145/1753846.1753855. 20

M. A. Rodriguez and P. Neubauer. Constructions from Dots and Lines. *Bulletin of the American Society for Information Science and Technology*, 36(6):35–41, 2010. DOI: 10.1002/bult.2010.1720360610. 37

R. E. Roth. An Empirically-Derived Taxonomy of Interaction Primitives for Interactive Cartography and Geovisualization. *IEEE Transactions on Visualization and Computer Graphics*, 19 (12):2356–2365, 2013. DOI: 10.1109/TVCG.2013.130. 27, 44, 73

R. Sadana and J. Stasko. Designing and Implementing an Interactive Scatterplot Visualization for a Tablet Computer. In *Proceedings of the Conference on Advanced Visual Interfaces (AVI)*, pages 265–272. ACM, 2014. DOI: 10.1145/2598153.2598163. 29

P. Saraiya, C. North, V. Lam, and K. Duca. An Insight-Based Longitudinal Study of Visual Analytics. *IEEE Transactions on Visualization and Computer Graphics*, 12(6):1511–1522, Nov 2006. DOI: 10.1109/TVCG.2006.85. 19

A. Satyanarayan, K. Wongsuphasawat, and J. Heer. Declarative Interaction Design for Data Visualization. In *Proceedings of the ACM Symposium on User Interface Software and Technology (UIST)*, pages 669–678. ACM, 2014. DOI: 10.1145/2642918.2647360. 72

H.-J. Schulz, T. Nocke, M. Heitzler, and H. Schumann. A Design Space of Visualization Tasks. *IEEE Transactions on Visualization and Computer Graphics*, 19(12):2366–2375, 2013a. DOI: 10.1109/TVCG.2013.120. 27

H.-J. Schulz, M. Streit, T. May, and C. Tominski. Towards a Characterization of Guidance in Visualization. Poster presentation, IEEE Conference on Information Visualization (InfoVis), 2013b. URL http://www.informatik.uni-rostock.de/~ct/pub_files/Schulz13Guidance.pdf. 66, 73

K. Sedig and P. Parsons. Interaction Design for Complex Cognitive Activities with Visual Representations: A Pattern-Based Approach. *AIS Transactions on Human-Computer Interaction*, 5 (2):84–133, 2013. 12, 27, 28, 73

K. Sedig, P. Parsons, and A. Babanski. Towards a Characterization of Interactivity in Visual Analytics. *Journal of Multimedia Processing and Technologies*, 3(1):12–28, 2012. 20

B. Shneiderman. Direct Manipulation: A Step Beyond Programming Languages. *IEEE Computer*, 16(8):57–69, 1983. DOI: 10.1109/MC.1983.1654471. 9, 28, 52

B. Shneiderman. Dynamic Queries for Visual Information Seeking. *IEEE Software*, 11(6): 70–77, 1994. DOI: 10.1109/52.329404. 11, 15, 30

B. Shneiderman. The Eyes Have It: A Task by Data Type Taxonomy for Information Visualizations. In *Proceedings of the IEEE Symposium on Visual Languages (VL)*, pages 336–343. IEEE Computer Society, 1996. 10, 27

B. Shneiderman and C. Plaisant. *Designing the User Interface: Strategies for Effective Human-Computer Interaction*. Addison Wesley, 5th edition, 2009. 8, 9

R. Spence. *Information Visualization: Design for Interaction*. Prentice-Hall, 2nd edition, 2007. 1, 9, 11, 12, 30, 37, 62, 71

M. Spindler, S. Stellmach, and R. Dachselt. PaperLens: Advanced Magic Lens Interaction Above the Tabletop. In *Proceedings of the International Conference on Interactive Tabletops and Surfaces (ITS)*, pages 69–76. ACM, 2009. DOI: 10.1145/1731903.1731920. 53

M. Spindler, C. Tominski, H. Schumann, and R. Dachselt. Tangible Views for Information Visualization. In *Proceedings of the International Conference on Interactive Tabletops and Surfaces (ITS)*, pages 157–166. ACM, 2010. DOI: 10.1145/1936652.1936684. 52, 54

M. Spindler, M. Martsch, and R. Dachselt. Going Beyond the Surface: Studying Multi-layer Interaction Above the Tabletop. In *Proceedings of the SIGCHI Conference Human Factors in Computing Systems (CHI)*, pages 1277–1286. ACM, 2012. DOI: 10.1145/2207676.2208583. 55

M. Spindler, M. Schuessler, M. Martsch, and R. Dachselt. Pinch-Drag-Flick vs. Spatial Input: Rethinking Zoom & Pan on Mobile Displays. In *Proceedings of the SIGCHI Conference Human Factors in Computing Systems (CHI)*, pages 1113–1122. ACM, 2014. DOI: 10.1145/2556288.2557028. 55

D. Stalling and H.-C. Hege. Fast and Resolution Independent Line Integral Convolution. In *Proceedings of the Annual Conference on Computer Graphics and Interactive Techniques (SIGGRAPH)*, pages 249–256. ACM, 1995. DOI: 10.1145/218380.218448. 30

C. D. Stolper, A. Perer, and D. Gotz. Progressive Visual Analytics: User-Driven Visual Exploration of In-Progress Analytics. *IEEE Transactions on Visualization and Computer Graphics*, 20 (12):1653–1662, 2014. DOI: 10.1109/TVCG.2014.2346574. 36

J. B. Strother, J. M. Ulijn, and Z. Fazal, editors. *Information Overload: An International Challenge for Professional Engineers and Technical Communicators*. Wiley, 2012. 1

R. Tamassia, editor. *Handbook of Graph Drawing and Visualization*. CRC Press, 2013. 37

D. Tang, C. Stolte, and R. Bosch. Design Choices When Architecting Visualizations. *Information Visualization*, 3(2):65–79, 2004. DOI: 10.1057/palgrave.ivs.9500067. 31

J. J. Thomas and K. A. Cook. *Illuminating the Path: The Research and Development Agenda for Visual Analytics*. IEEE Computer Society, 2005. 10, 18, 19

C. Tominski. Event-Based Concepts for User-Driven Visualization. *Information Visualization*, 10(1):65–81, 2011. DOI: 10.1057/ivs.2009.32. 60, 61

C. Tominski. Foldable Visualization. Interactive prototype, 2012. URL http://goo.gl/LwREL. Retrieved: June, 2015. 48

C. Tominski, J. Abello, and H. Schumann. Axes-Based Visualizations with Radial Layouts. In *Proceedings of the ACM Symposium on Applied Computing (SAC)*, pages 1242–1247. ACM, 2004. DOI: 10.1145/967900.968153. 35, 61

C. Tominski, G. Fuchs, and H. Schumann. Task-Driven Color Coding. In *Proceedings of the International Conference Information Visualisation (IV)*, pages 373–380. IEEE Computer Society, 2008. DOI: 10.1109/IV.2008.24. 27

C. Tominski, J. Abello, and H. Schumann. CGV – An Interactive Graph Visualization System. *Computers & Graphics*, 33(6):660–678, 2009. DOI: 10.1016/j.cag.2009.06.002. 35, 37, 40

C. Tominski, J. F. Donges, and T. Nocke. Information Visualization in Climate Research. In *Proceedings of the International Conference Information Visualisation (IV)*, pages 298–305. IEEE Computer Society, 2011a. DOI: 10.1109/IV.2011.12. 20

C. Tominski, H. Schumann, M. Spindler, and R. Dachselt. Towards Utilizing Novel Interactive Displays for Information Visualization. In *Proceedings of the Workshop on Data Exploration for Interactive Surfaces (DEXIS)*. HAL - Inria Open Archive, 2011b. URL http://hal.inria.fr/hal-00659469. 29

C. Tominski, C. Forsell, and J. Johansson. Interaction Support for Visual Comparison Inspired by Natural Behavior. *IEEE Transactions on Visualization and Computer Graphics*, 18(12):2719–2728, 2012a. DOI: 10.1109/TVCG.2012.237. 46, 48

C. Tominski, H. Schumann, G. Andrienko, and N. Andrienko. Stacking-Based Visualization of Trajectory Attribute Data. *IEEE Transactions on Visualization and Computer Graphics*, 18 (12):2565–2574, 2012b. DOI: 10.1109/TVCG.2012.265. 40, 42, 44

C. Tominski, H. Schumann, G. Andrienko, and N. Andrienko. Stacking-Based Visualization of Trajectory Attribute Data. Interactive prototype, 2012c. URL `http://goo.gl/wIC1k`. Retrieved: June, 2015. 44

C. Tominski, S. Gladisch, U. Kister, R. Dachselt, and H. Schumann. A Survey on Interactive Lenses in Visualization. In *EuroVis State-of-the-Art Reports*, pages 43–62. Eurographics Association, 2014. DOI: 10.2312/eurovisstar.20141172. 44, 71

C. Tominski, A. Berdowski, C. Tessnow, and S. Plath. iGraph.js – Interaction for Graph Visualization. Interactive prototype, 2015. URL `http://www.informatik.uni-rostock.de/~ct/software/iGraph.js/`. Retrieved: June, 2015. 40

M. Tory and T. Möller. Human Factors in Visualization Research. *IEEE Transactions on Visualization and Computer Graphics*, 10(1):72–84, 2004. DOI: 10.1109/TVCG.2004.1260759. 29

L. A. Treinish. Task-Specific Visualization Design. *IEEE Computer Graphics and Applications*, 19(5):72–77, 1999. DOI: 10.1109/38.788803. 27

B. Ullmer, H. Ishii, and R. J. K. Jacob. Tangible Query Interfaces: Physically Constrained Tokens for Manipulating Database Queries. In *Proceedings of the TC13 IFIP International Conference on Human-Computer Interaction (INTERACT)*, pages 279–286. IOS Press, 2003. 55

A. Valli. The Design of Natural Interaction. *Multimedia Tools and Applications*, 38(3):295–305, 2008. DOI: 10.1007/s11042-007-0190-z. 9, 30, 48

A. van Dam. Post-WIMP User Interfaces. *Communications of the ACM*, 40(2):63–67, 1997. DOI: 10.1145/253671.253708. 9

S. van den Elzen and J. J. van Wijk. Multivariate Network Exploration and Presentation: From Detail to Overview via Selections and Aggregations. *IEEE Transactions on Visualization and Computer Graphics*, 20(12):2310–2319, 2014. DOI: 10.1109/TVCG.2014.2346441. 10

F. van Ham and A. Perer. Search, Show Context, Expand on Demand: Supporting Large Graph Exploration with Degree-of-Interest. *IEEE Transactions on Visualization and Computer Graphics*, 15(6):953–960, 2009. DOI: 10.1109/TVCG.2009.108. 65

F. van Ham and J. J. van Wijk. Interactive Visualization of Small World Graphs. In *Proceedings of the IEEE Symposium Information Visualization (InfoVis)*, pages 199–206. IEEE Computer Society, 2004. DOI: 10.1109/INFVIS.2004.43. 40

J. J. van Wijk. The Value of Visualization. In *Proceedings of the IEEE Visualization Conference (Vis)*, pages 79–86. IEEE Computer Society, 2005. DOI: 10.1109/VIS.2005.102. 48

J. J. van Wijk. Views on Visualization. *IEEE Transactions on Visualization and Computer Graphics*, 12(4):421–433, 2006. DOI: 10.1109/TVCG.2006.80. 6

J. J. van Wijk and W. A. A. Nuij. A Model for Smooth Viewing and Navigation of Large 2D Information Spaces. *IEEE Transactions on Visualization and Computer Graphics*, 10(4):447–458, 2004. DOI: 10.1109/TVCG.2004.1. 14, 38

B. Victor. Magic Ink – Information Software and the Graphical Interface, 2006. URL http://worrydream.com/MagicInk. Retrieved: June, 2015. 21

S. Voida, M. Tobiasz, J. Stromer, P. Isenberg, and M. S. T. Carpendale. Getting Practical with Interactive Tabletop Displays: Designing for Dense Data, "Fat Fingers", Diverse Interactions, and Face-To-Face Collaboration. In *Proceedings of the International Conference on Interactive Tabletops and Surfaces (ITS)*, pages 109–116. ACM, 2009. DOI: 10.1145/1731903.1731926. 29, 52

T. von Landesberger, A. Kuijper, T. Schreck, J. Kohlhammer, J. J. van Wijk, J.-D. Fekete, and D. W. Fellner. Visual Analysis of Large Graphs: State-of-the-Art and Future Research Challenges. *Computer Graphics Forum*, 30(6):1719–1749, 2011. DOI: 10.1111/j.1467-8659.2011.01898.x. 37, 40

M. Ward and J. Yang. Interaction Spaces in Data and Information Visualization. In *Proceedings of the Joint Eurographics – IEEE TCVG Symposium on Visualization (VisSym)*, pages 137–146. Eurographics Association, 2004. URL http://diglib.eg.org/EG/DL/WS/VisSym/VisSym04/137-146.pdf. 12

M. O. Ward, G. Grinstein, and D. Keim. *Interactive Data Visualization: Foundations, Techniques, and Applications*. A K Peters/CRC Press, 2010. 12, 27

C. Ware. *Visual Thinking for Design*. Morgan Kaufmann, 2008. 5

C. Ware. *Information Visualization: Perception for Design*. Morgan Kaufmann, 3rd edition, 2012. 1, 11, 12, 36, 45

C. Ware, R. Arsenault, M. Plumlee, and D. Wiley. Visualizing the Underwater Behavior of Humpback Whales. *Computer Graphics and Applications*, 26(4):14–18, 2006. DOI: 10.1109/MCG.2006.93. 40

C. Weaver. Building Highly-Coordinated Visualizations in Improvise. In *Proceedings of the IEEE Symposium Information Visualization (InfoVis)*, pages 159–166. IEEE Computer Society, 2004. DOI: 10.1109/INFVIS.2004.12. 22

C. E. Weaver and M. Livny. Improving Visualization Interactivity in Java. In *Proceedings of the Conference on Visualization and Data Analysis (VDA)*, pages 62–72. SPIE, 2000. DOI: 10.1117/12.378919. 36

P. Wegner. Why Interaction Is More Powerful Than Algorithms. *Communications of the ACM*, 40(5):80–91, 1997. DOI: 10.1145/253769.253801. 20

A. Wiebel, F. M. Vos, D. Foerster, and H.-C. Hege. WYSIWYP: What You See Is What You Pick. *IEEE Transactions on Visualization and Computer Graphics*, 18(12):2236–2244, 2012. DOI: 10.1109/TVCG.2012.292. 17

D. Wigdor and D. Wixon. *Brave NUI World: Designing Natural User Interfaces for Touch and Gesture*. Morgan Kaufmann, 2011. 30

L. Wilkinson. *The Grammar of Graphics*. Springer, 2nd edition, 2005. DOI: 10.1007/0-387-28695-0. 74

N. Willems, H. van de Wetering, and J. J. van Wijk. Visualization of Vessel Movements. *Computer Graphics Forum*, 28(3):959–966, 2009. DOI: 10.1111/j.1467-8659.2009.01440.x. 40

G. Wills. *Visualizing Time*. Springer, 2012. DOI: 10.1007/978-0-387-77907-2. 40

G. J. Wills. Selection: 524,288 Ways to Say "This is Interesting". In *Proceedings of the IEEE Symposium Information Visualization (InfoVis)*, pages 54–60. IEEE Computer Society, 1996. DOI: 10.1109/INFVIS.1996.559216. 15

N. Wong, M. S. T. Carpendale, and S. Greenberg. EdgeLens: An Interactive Method for Managing Edge Congestion in Graphs. In *Proceedings of the IEEE Symposium Information Visualization (InfoVis)*, pages 51–58. IEEE Computer Society, 2003. DOI: 10.1109/INFVIS.2003.1249008. 40

M. Wybrow, N. Elmqvist, J.-D. Fekete, T. von Landesberger, J. J. van Wijk, and B. Zimmer. Interaction in the Visualization of Multivariate Networks. In A. Kerren, H. C. Purchase, and M. O. Ward, editors, *Multivariate Network Visualization*, volume 8380 of *Lecture Notes in Computer Science*, pages 97–125. Springer, 2014. DOI: 10.1007/978-3-319-06793-3_6. 27, 37, 40

J. S. Yi, Y. ah Kang, J. T. Stasko, and J. A. Jacko. Toward a Deeper Understanding of the Role of Interaction in Information Visualization. *IEEE Transactions on Visualization and Computer Graphics*, 13(6):1224–1231, 2007. DOI: 10.1109/TVCG.2007.70515. 2, 12, 27, 28

F. Ying, P. Mooney, P. Corcoran, and A. C. Winstanley. Dynamic Visualization of Geospatial Data on Small Screen Mobile Devices. In G. Gartner and F. Ortag, editors, *Advances in Location-Based Services*, pages 77–90. Springer, 2012. DOI: 10.1007/978-3-642-24198-7_5. 28

B. Yost and C. North. The Perceptual Scalability of Visualization. *IEEE Transactions on Visualization and Computer Graphics*, 12(5):837–844, 2006. DOI: 10.1109/TVCG.2006.184. 28, 58

L. Yu, K. Efstathiou, P. Isenberg, and T. Isenberg. Efficient Structure-Aware Selection Techniques for 3D Point Cloud Visualizations with 2DOF Input. *IEEE Transactions on Visualization and Computer Graphics*, 18(12):2245–2254, 2012. DOI: 10.1109/TVCG.2012.217. 17, 29

J. Zhao, F. Chevalier, and R. Balakrishnan. KronoMiner: Using Multi-foci Navigation for the Visual Exploration of Time-series Data. In *Proceedings of the SIGCHI Conference Human Factors in Computing Systems (CHI)*, pages 1737–1746. ACM, 2011. DOI: 10.1145/1978942.1979195. 27

M. X. Zhou and S. K. Feiner. Visual Task Characterization for Automated Visual Discourse Synthesis. In *Proceedings of the SIGCHI Conference Human Factors in Computing Systems (CHI)*, pages 392–399. ACM, 1998. DOI: 10.1145/274644.274698. 27

M. Zinsmaier, U. Brandes, O. Deussen, and H. Strobelt. Interactive Level-of-Detail Rendering of Large Graphs. *IEEE Transactions on Visualization and Computer Graphics*, 18(12):2486–2495, 2012. DOI: 10.1109/TVCG.2012.238. 36

E. Zudilova-Seinstra, T. Adriaansen, and R. van Liere. *Trends in Interactive Visualization: State-of-the-Art Survey*. Springer, 2009. 17

Author's Biography

CHRISTIAN TOMINSKI

Christian Tominski received a diploma and doctoral degree from the University of Rostock, Germany, in 2002 and 2006, respectively. He is a senior researcher and lecturer at the Institute for Computer Science at the University of Rostock. His research interests are in visualization and visual analytics. He is particularly interested in the role of interaction for visual data exploration and analysis. He worked on utilizing novel display and interaction devices for interactive visualization and on integrating automatic methods to assist the visualization. Christian has authored and co-authored more than 50 academic publications on new visualization approaches and interaction methods. He also developed a number of visualization tools for time-oriented data, spatio-temporal data, movement data, and graph data. More about Christian's research and demos of visualization prototypes can be found on his website at: `http://www.informatik.uni-rostock.de/~ct`.

Printed in the United States
by Baker & Taylor Publisher Services